Studies in Fuzziness and Soft Computing

Volume 413

The series "Studies in Fuzziness and Soft Computing" contains publications on various topics in the area of soft computing, which include fuzzy sets, rough sets, neural networks, evolutionary computation, probabilistic and evidential reasoning, multi-valued logic, and related fields. The publications within "Studies in Fuzziness and Soft Computing" are primarily monographs and edited volumes. They cover significant recent developments in the field, both of a foundational and applicable character. An important feature of the series is its short publication time and world-wide distribution. This permits a rapid and broad dissemination of research results.

Indexed by SCOPUS, DBLP, WTI Frankfurt eG, zbMATH, SCImago.

All books published in the series are submitted for consideration in Web of Science.

More information about this series at https://link.springer.com/bookseries/2941

Deepak Gupta · Aditya Khamparia ·
Ashish Khanna · Oscar Castillo
Editors

Soft Computing for Data Analytics, Classification Model, and Control

Springer

Editors
Deepak Gupta
Department of Computer Science
Engineering
Maharaja Agrasen Institute of Technology
Rohini, Delhi, India

Aditya Khamparia
Department of Computer Science
Babasaheb Bhimrao Ambedkar (Central
University), Satellite Centre
Amethi, Uttar Pradesh, India

Ashish Khanna
Maharaja Agrasen Institute of Technology
Rohini, Delhi, India

Oscar Castillo
Tijuana Institute of Technology
Tijuana, Baja California, Mexico

ISSN 1434-9922 ISSN 1860-0808 (electronic)
Studies in Fuzziness and Soft Computing
ISBN 978-3-030-92025-8 ISBN 978-3-030-92026-5 (eBook)
https://doi.org/10.1007/978-3-030-92026-5

This Springer imprint is published by the registered company Springer Nature Switzerland AG
The registered company address is: Gewerbestrasse 11, 6330 Cham, Switzerland

Preface

This book presents a set of soft computing approaches and their application in data analytics, classification model, and control. The basics of fuzzy logic implementation to advanced hybrid fuzzy-driven optimization methods has been covered in the book. The various soft computing techniques including fuzzy logic, rough sets, neutrosophic sets, type-2 fuzzy logic, neural networks, generative adversarial networks, and evolutionary computation have been discussed, and they are used on variety of applications including data analytics, classification model, and control.

The book is divided into two thematic parts. The first thematic section covers the various soft computing approaches for text classification and data analysis, while the second section focuses on the fuzzy-driven optimization methods for the control systems.

Part One: Fuzzy Logic Implication and Usage in Text Classification and Medical Data Analytics

Medical imaging is at the heart of many of today's improved diagnostic and treatment technologies. The preprocessing tasks of filtering, sharpening, and focusing an image are discussed in this part. This part covers the implementation of the fuzzy logic method in each of the fundamental steps of medical disease diagnosis and classification analysis. This part includes four chapters. The first chapter by Sujatha Krishanmoorthy et al. discusses a new sentiment and emotion classification model based on the fuzzy rough set nearest neighbor (FRSNN) and krill herd (KH) algorithm. The proposed model outperforms the other existing methods in literature. Next chapter by Sujatha Krishanmoorthy et al. describes the fuzzy wavelet neural network (FWNN) with social spider optimization (SSO) algorithm to recognize the medical data. The experimental validation of the SSO-FWNN model takes place using three medical benchmark datasets, viz. breast cancer, heart disease, and hepatitis dataset. The third chapter of this section by Denis A. Pustokhin et al. describes a new diabetes disease diagnosis model using the radial basis function (RBF) with fuzzy-based gravitational search algorithm (FGSA). The proposed FGSA-RBF model is used to determine the optimal structure of the RBF network, and it outperforms the other existing related models in the literature. The last chapter in this section by V. Gomathy et al. focuses on rule-based extraction methods for data classification. An optimized oppositional-based elephant herd algorithm (OEHA) has been used to generate the rules for neutrosophic set (NS).

Part Two: Fuzzy Driven Optimization Methods and Control Systems

This part presents novel fuzzy-driven optimization methods for various applications such as NP-hard problems, chaotic systems, power and control systems. This part covers five chapters. The first chapter by Tien-Loc Le et al. presents a self-evolving interval type-2 fuzzy neural network using a function-link network and asymmetric membership functions, which are used to synchronize the chaotic systems. The self-evolving algorithm is applied for constructing the network structure autonomously. The function-link network and a symmetric membership function are applied to enhance the learning capability of the proposed network. The second chapter by Shubham Agrawal et al. proposes a genetic algorithm-based approach for optimizing the hyperparameters. The proposed approach was tested on two standard datasets representing two different languages: English and Hindi. Next chapter by Jose A. Barrios-Gomez et al. describes a data-driven fuzzy c-means technique for power system frequency generation, which outperforms over different existing machine learning methods and introduces better accuracy for the system. Fourth chapter by Palash Dutta and Gourangajit Borah proposes a novel similarity measure based on newly defined expected values and variances of fuzzy triangular numbers and describes an application to group decision making. The proposed similarity measure approach is based on defining two expected values and their variances for the left and proper reference functions of a fuzzy triangular numbers. The last chapter of this book by Abedin et al. describes a new generative adversarial net (GAN)-based architecture, which can generate printed characters from handwritten Bangla characters. The GAN model works on the principle of an image-to-image translation based on image mapping to output. Due to variations in data distribution, generated printed characters can train an optical character recognition model. The loss of function is not required to generate printed characters from handwritten characters.

Rohini, India — Deepak Gupta
Amethi, India — Aditya Khamparia
Rohini, India — Ashish Khanna
Tijuana, Mexico — Oscar Castillo

Contents

Fuzzy Logic Implication and Usage in Text Classification and Medical Data Analytics

An Optimization of Fuzzy Rough Set Nearest Neighbor Classification Model Using Krill Herd Algorithm for Sentiment Text Analytics 3
Sujatha Krishanmoorthy, Yunsheng Chai, Yewei Du, Xiangnan Zhou, Huimin Zhang, Hongyu Liu, and Changjiang Zhang

Fuzzy Wavelet Neural Network with Social Spider Optimization Algorithm for Pattern Recognition in Medical Domain 23
Sujatha Krishanmoorthy, S. Selvakumar, Veeraraghavan Vishnu Priya, Surapaneni Krishna Mohan, and K. Sabithabanu

Fuzzy with Gravitational Search Algorithm Tuned Radial Basis Function Network for Medical Disease Diagnosis and Classification Model 41
Denis A. Pustokhin, Irina V. Pustokhina, M. Ilayaraja, and K. Shankar

Optimal Neutrosophic Rules Based Feature Extraction for Data Classification Using Deep Learning Model 57
V. Gomathy, T. Jayasankar, M. Rajaram, E. Anna Devi, and S. Priyadharshini

Fuzzy Driven Optimization Methods and Control Systems

Self-Evolving Interval Type-2 Fuzzy Neural Network Design for the Synchronization of Chaotic Systems 83
Tien-Loc Le, Chih-Min Lin, Tuan-Tu Huynh, and Sung Kyung Hong

Categorizing Relations via Semi-supervised Learning Using a Hybrid Tolerance Rough Sets and Genetic Algorithm Approach 103
Shubham Agrawal, Rashad Ahmed, M. Anand Kumar, and Sheela Ramanna

Data-Driven Fuzzy C-Means Equivalent Turbine-Governor for Power System Frequency Response 117
Jose A. Barrios-Gomez, Francisco Sanchez, Francisco Gonzalez-Longatt, Gianfranco Claudio, Alberto Cavazos, Harold R. Chamorro, and Wilmar Martinez

Multicriteria Group Decision Making Using a Novel Similarity Measure for Triangular Fuzzy Numbers Based on Their Newly Defined Expected Values and Variances 137
Palash Dutta and Gourangajit Borah

Bangla Printed Character Generation from Handwritten Character Using GAN 153
Md. Min-ha-zul Abedin, Tapotosh Ghosh, Tazqia Mehrub, and Mohammad Abu Yousuf

Fuzzy Logic Implication and Usage in Text Classification and Medical Data Analytics

An Optimization of Fuzzy Rough Set Nearest Neighbor Classification Model Using Krill Herd Algorithm for Sentiment Text Analytics

Sujatha Krishanmoorthy, Yunsheng Chai, Yewei Du, Xiangnan Zhou, Huimin Zhang, Hongyu Liu, and Changjiang Zhang

Abstract In recent years, social networking sites like Twitter and Facebook are more popular through which emotions and thoughts of its users are shared. Detection and classification of emotions, expressed in social media content, is beneficial for several applications in the domains like e-commerce, politics, social welfare and so on. Several works have been conducted earlier focusing sentiment and emotion analysis. These works primarily concentrated on single-label classification, leaving beside the correlation among different emotions expressed by an individual. In this view, the current research article presents a new sentiment and emotion classification model using Fuzzy Rough Set Nearest Neighbor (FRSNN) with Krill Herd (KH) algorithm i.e., FRSNN-KH. The proposed FRSNN-KH algorithm involves preprocessing, feature extraction and classification. Initially, preprocessing is executed to remove the unwanted words present in the tweet. Next, the features are extracted from the pre-processed tweet by following Bag of Words (BoW) model. Afterwards, FRSNN-based classification process is carried out to segregate the instances under different class labels. Finally, soft computing-based KH algorithm is applied to optimize the

S. Krishanmoorthy (✉) · Y. Chai · Y. Du · X. Zhou · H. Zhang · H. Liu
Department of Computer Science, Wenzhou Kean University, Zhejiang Province, Wenzhou, China
e-mail: sujatha@wku.edu.cn

Y. Chai
e-mail: Chaiyu@kean.edu

Y. Du
e-mail: Duy@kean.edu

X. Zhou
e-mail: zhouxian@kean.edu

H. Zhang
e-mail: zhanghui@kean.edu

H. Liu
e-mail: liuhon@kean.edu

C. Zhang
Division of Science and Technology, Hongkong Baptist University, UIC, Zhuhai, China
e-mail: changjiangzhang@uic.edu.hk

D. Gupta et al. (eds.), *Soft Computing for Data Analytics, Classification Model, and Control*, Studies in Fuzziness and Soft Computing 413,
https://doi.org/10.1007/978-3-030-92026-5_1

rule generation sets by FRSNN model. The presented FRSNN-KH algorithm was experimentally validated using benchmark dataset. The simulation outcome inferred the goodness of the proposed FRSNN-KH algorithm over compared methods under several dimensions.

Keywords Classification · Emotions · Fuzzy rough set theory · Sentiment analysis · Soft computing

1 Introduction

Sentiment Analysis (SA) study established that there is a frequent development of web by means of volume, velocity and variety of opinion-rich data available in Internet. SA is a tendency for real-time functions and serves as decision support system. It further submits the required data to analysts. Specifically, big data analytics is an effective technique which leverages highly unstructured data and makes a significant data analysis with the help of contemporary tools. In addition, text-mining methods describe the transformation and replacement of unstructured data into structured one for the purpose of Knowledge Discovery (KD). Extensive studies have been carried out to find the best text mining process. Here, SA is referred to as a general text classification operation [1]. It studies the people's suggestions, approach, and sentiments towards an object or any kind of actions. The people's responses might be views and suggestion polarities such as positive, negative or neutral for textual context, accessible by social medium.

In typical SA process, few steps are followed such as data collection, Feature Selection (FS) [2], sentiment classification and sentiment polarity detection. Figure 1 shows the general structure of SA. An effective FS process is said to be a complicated one. It plays a vital role in computing accurate sentiment classification. Additionally, enhanced dimensionality, difficulty, and fuzziness in user-based Twitter data promote the requirement for best sentiment classification models. There are few prominent studies conducted earlier in search of novel concepts that can handle uncertainty, inaccuracy, estimation, incomplete truth and fuzziness. These concepts are demanded for options to enable the replication of human intelligence for personalized as well as common results. Soft Computing (SC) is another application which tends to integrate fresh processing approaches that reflect the consciousness and cognition in diverse dimensions, a knowledge acquired through experience; it can be universalized into applications that lack direct knowledge; and by parallel computer structures which accelerates the biological task. Soft computing is capable of performing rapid mapping of inputs to outputs when compared with inherent serial analytical presentations.

Followed by, a programmatic study on "Sentiment Analysis of Twitter Using SC Methods" has been presented in the literature. The main aim of this model is to collect empirical witness, examine the results from previous studies to provide better outcomes than previous studies, find the gaps in recent search and offer future

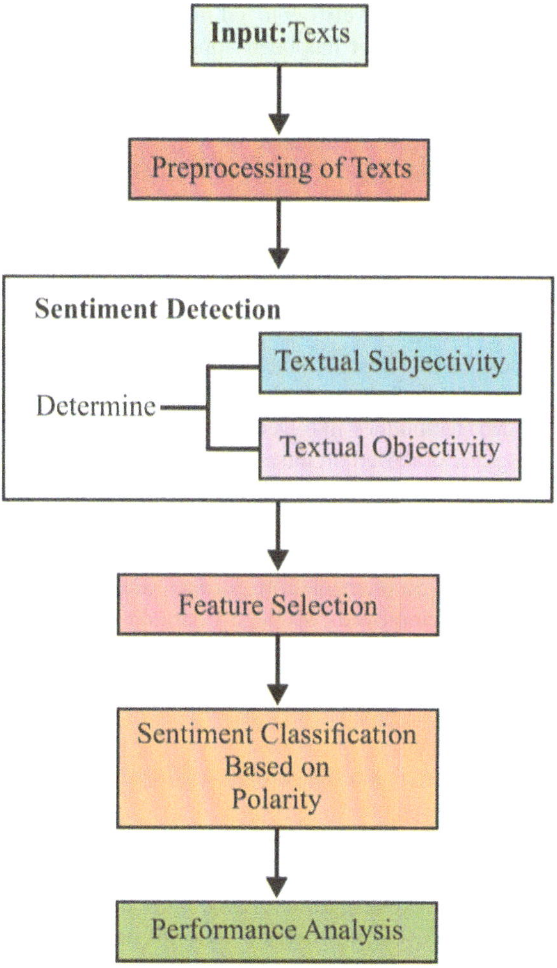

Fig. 1 General structure of SA

prospects with respect to the methods developed so far. SC models are assumed to be optimizing approaches which tend to deploy tedious real-time issues that can accomplish strong and minimum cost schemes. Such models are classified into five classes such as Machine Learning (ML), Neural Networks (NN), Evolutionary Computation, Fuzzy Logic (FL) and Probabilistic Reasoning (PR): Bayesian Networks (Bayesian probability).

Sentiments or emotions are mainly used to convey the feelings and opinions among human beings. Online social media, like Twitter and Facebook (FB), are offered in different languages depending on the people. Recently, an update has been given in which the users can make short interactions regarding facts, opinions, emotions, and its intensities on diverse kinds of themes. Authors are fascinated towards the determination of emotions in social media content using natural language processing.

It has been extensively deployed in commerce, public health, social welfare, and so on. For example, it has been applied in public health, public opinion prediction regarding political movements, brand consultation, and stock market observation. SA is defined as the estimation of attitude for the topic under study. The attitude may be either positive or negative or sometimes it may be even an emotional state like happy, sad and anger [3]. Currently, multi-label classification issues have achieved applicable results from a vast range of applications, such as text, scene and video classification, as well as bioinformatics [4]. On the contrary, in classical single-label classification problem, the sample is related with a single label from certain labels. However, in case of multi-label classification problem, a sample is combined with a subset of labels.

Several ML models have been presented for conventional emotion classification and multi-label emotion classification so far [5–8]. Al-Smadi et al. [9] presented two ML techniques like Deep Recurrent Neural Network (DRNN) and SVM. The researchers focused on the assessment of three different operations like aspect class examination, aspect Opinion Target Expression (OTE), and aspect sentiment polarity identification. When compared, the attained results implied that DRNN is an effective performer compared to SVM. Thus, DRNN is said to be robust and rapid in terms of implementation during training and testing. Furthermore, Al-Smadi et al. [10] employed two deep Learning Long Short-Term Memory (LSTM) NN for aspect-based SA. Initially, a character-level bidirectional LSTM was applied in aspect OTE extraction with a definite classification. Alternatively, an aspect-based LTSM was presented in this study for aspect sentiment orientation classification. The final outcome shows that the aspect-OTE extraction as well as orientation classifier function were better when related to baseline study.

In line with this, Al-Ayyoub et al. [11] employed a supervised ML model to extract different aspects and categorize the sentiments for hotels' Arabic reviews. The model has three major operations such as finding the aspect classes, obtaining the opinion targets, and exploring the sentiment polarity. The simulation outcome pointed that the newly-developed method performs well using a similar dataset, compared to other applications. Besides, García-Pablos et al. [12] provided W2VLDA, an aspect sentiment classifier model which needs minimum examination and no necessity of specific languages. It is capable of differentiating opinion words from aspect terms in an unsupervised manner. The model's efficiency was estimated by applying SemEval 2016 dataset. The analysis implied competing outcome for diverse languages. Likewise, Dragoni et al. [13] deployed an aspect-based unsupervised model for opinion tracking which maintains data visualization.

Rathan et al. [14] implied an ontology-based SA method for 'Smart phones' in case of tweets and messages. Here, the model was designed with diverse features like spelling correction, emoji and sentiment detection. Lexicon-driven technology was utilized for automatic training of data labels. A 'Smartphone'-based application, loaded with SVM, was used to enhance the classification accuracy. Cambria et al. [15] presented an ensemble model of symbolic and sub-symbolic AI. In this model, LSTM network was used to find the verb-noun primitives through lexical replacement and connect to common-sense paradigm using 3-level knowledge representation for

SA named 'SenticNet'. Alternatively, Xianghua et al. [16] recommended an unsupervised method to estimate the aspects and emotions in Chinese social reviews. The developers applied Latent Dirichlet Allotiment (LDA) in social reviews to explore the multi-aspect global sections.

Chen et al. [17] boosted LDA to arrive at Automatic Knowledge LDA with complete automation instead of applying the previous domains to find novel aspects. This model was capable of producing aspects and resolving the issues relevant to inferior knowledge through application and improvement of Gibbs sampler approach. Poria et al. [18] arranged an unsupervised rule centric technique to get explicit aspects and Implicit Aspects Clues (IAC) from goods and restaurant reviews. In this model, the researchers applied unsupervised rule centric technique to explore IACs in review. Then, it is divided into aspects with common-sense knowledge and dependency structure of a sentence, under the application of WordNet and SenticNet.

Pham and Le [19] projected a multi-layer structure to point the hotels reviews of users. Learning methods like word embeddings and compositional vectors were employed in the study to collect words, sentences, aspect-based graphs, etc. These elements were used to arrive at productive knowledge. Such representations were integrated with NN to develop a method that can predict the entire hotel reviews at the time of extracting aspects' ratings and weights. Liu et al. [20] concatenated supervised and unsupervised models in which automated rule-based approaches were used to enhance double propagation. Hence, double propagation considers those opinion words with a target. There is a syntactic relationship between the opinion word and a target with similar sentence. Ma et al. [21] integrated LSTM network with hierarchical process. The simulation outcome shows that the integration of planned attention design and Sentic LSTM performs quite well than aspect-based SA process. If the application is composed of autonomous parameters to define the link between aspects and corresponding opinion representations, it is capable of capturing aspect-driven emotions with few data requirement.

The authors, in current study, presented an efficient sentiment and emotion classification model using FRSNN and KH Algorithm, named as FRSNN-KH. The presented model involves preprocessing, BoW-based feature extraction, FRSNN-based classification and KH-based rule set optimization. The application of soft computing-based KH algorithm helps in the optimization of number of rules generated by FRSNN model. Further, it assists in improving the classification performance. The performance of the presented FRSNN-KH algorithm was validated using benchmark dataset and the results were assessed under several aspects.

2 The Proposed FRSNN-KH Model

Figure 2 shows the overall working principle of FRSNN-KH approach. The figure reveals that the input data undergoes preprocessing to remove the unwanted words present in the tweet. Followed by, BoW method is applied as a feature extractor, to

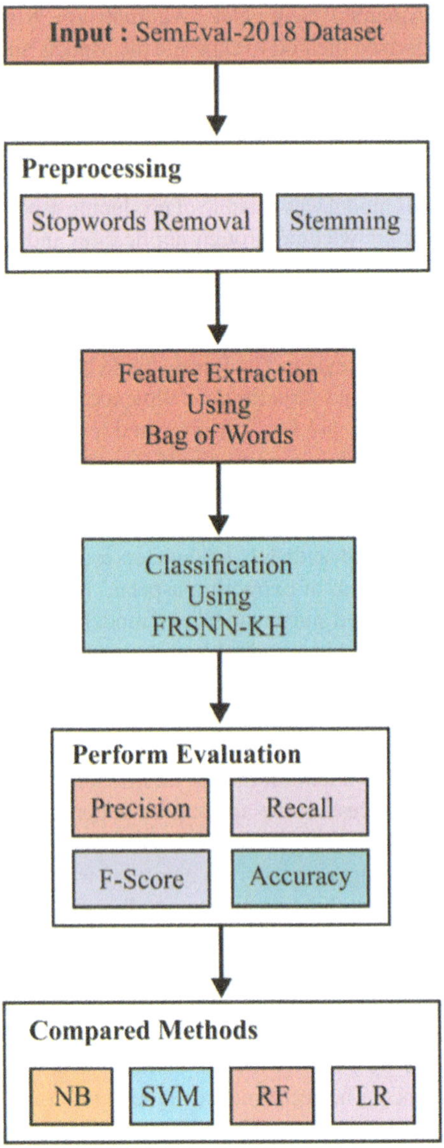

Fig. 2 Block diagram of FRSNN-KH algorithm

acquire the useful features from pre-processed tweet. Then, FRSNN-based classification is carried out to identify and segregate the instances under different class labels. Finally, soft computing-based KH algorithm is applied to optimize the rule generation sets by FRSNN model.

2.1 Fuzzy-Rough Set Theory

A study based on hybridization of fuzzy sets and rough sets was presented with maximum efficiency. It mainly concentrated on fuzzifying the equations for lower and upper approximations [22]. The study followed two basic principles as provided below:

- Set A is generalized to fuzzy set in X that enables the objects which come under the class of applied degrees.
- Instead of using objects with indiscernibility, it is better to estimate the appropriate quality. Finally, the objects are classified into classes or granules along with 'soft' boundaries according to similarity measure. In line with this, sudden transactions occur among these classes by moderate ones with the activation of a component that possess massive number of classes.

In particular, approximate equality is assessed among the objects with frequent attribute values by means of fuzzy relation R in X. This has been allocated for every object based on its similarity value. Basically, R is a minimum fuzzy tolerance formula in which the fuzzy set A of X is provided with lower and upper approximations of A alike how R is manufactured in various points. A common description as given herewith

$$(R \downarrow A)(p) = \underset{q \in X}{inf \ell} (R(p,q), A(q)) \tag{1}$$

$$(R \uparrow A)(p) = \underset{q \in X}{\sup \mathcal{T}} (R(p,q), A(q)) \tag{2}$$

where, ℓ implies an implicator and $\mathcal{T}$, a t-norm. Here, A defines a set and R denotes an equivalent relation in X. Fortunately, it is an effective advantage for specific complexities. For this purpose, the method of Vaguely Quantified Rough Sets (VQRS) is deployed in this study. It applies the linguistic quantifiers like 'most' and 'some' which are against traditional crisp quantifiers such as 'all' and 'at least one' since the latter selects the class of objects to both lower and upper approximations. A couple of fuzzy quantifiers (Q_u, Q_j) is applied which makes the 'most' and 'some', with lower and upper approximations of A by R as described herewith.

$$(R \downarrow^{Q_\mu} A)(q) = Q_u\left(\frac{|Rq \cap A|}{|Rq|}\right) = Q_u\left(\frac{\sum_{p \in X} \min(R(p,q), A(p))}{\sum_{p \in X} R(p,q)}\right) \tag{3}$$

$$(R \uparrow^{Q} A)(q) = Q_j\left(\frac{|Rq \cap A|}{|Rq|}\right) = Q_j\left(\frac{\sum_{p \in X} \min(R(p,q), A(p))}{\sum_{p \in X} R(p,q)}\right) \tag{4}$$

The intersection of a fuzzy set is represented by min t-norm. The instances of fuzzy quantifiers can be generated with respect to given parameterized equation, for $0 \leq \alpha < \beta \leq$ and p in [0, 1],

$$Q_{!\alpha,\beta)}(p) = \begin{cases} 0, & p \leq \alpha \\ \dfrac{2(p-\alpha)^2}{\beta-\alpha} & \alpha \leq p \leq \dfrac{\alpha+\beta}{2} \\ 1 - \dfrac{2(p-\alpha)^2}{(\beta-\alpha)^2} & \dfrac{\alpha+\beta}{2} \leq p \leq \beta \\ 1, & \beta \leq p \end{cases} \tag{5}$$

Here, $Q_{(0.1,0.6)}$ and $Q_{(0.2,1)}$ are applied to show vague quantifiers (VQ) '*some*' and '*most*' simultaneously. The vital difference between (8) and (9) is that the VQRS approximations do not expand the conventional FR approximations. Here, A and R are crisp and are represented herewith.

$$Q_{>p_l}(p) = \begin{cases} 0, & p \leq p_l \\ 1, & p > p_l \end{cases} \quad Q_{\geq p_u}(p) = \begin{cases} 0, & p < p_u \\ 1, & p \geq p_u \end{cases} \tag{6}$$

With $0 \leq p_l < p_u \leq 1$ being applied as quantifiers, Ziaroko's variable precision rough set model is recovered and it applies the following equation.

$$Q_{\exists}(p) = \begin{cases} 0, & p = 0 \\ 1, & p > 0 \end{cases} \quad Q_{\forall}(p) = \begin{cases} 0, & p < 1 \\ 1, & p = 1 \end{cases} \tag{7}$$

Pawlak's standard rough set method is applied in case of VQRS model in which R is considered as a crisp equivalence formula. Then, VQRS approach resolves the noisy data using a novel perception: it acquires the elasticity of VPRS to solve the classifier errors, releases the membership constraints for lower approximation, fix the upper approximation and fuzzy sets to express incomplete constraint convenience. This method was applied to perform the FS process.

2.2 *Fuzzy-Rough Set Nearest Neighbours (FRSNN)-Based Classification*

In this work, FRSNN method was presented in which NN was used to develop fuzzy lower and upper approximations of decision categories. Further, the test samples were divided according to the membership of such approximations. This technology depends upon fuzzy tolerance relation, R. Here, R is developed using the group of conditional attributes A, R as shown by,

$$R(p, q) = \min_{a \in \mathbb{A}} R_0(p, q) \tag{8}$$

where $R_a(p, q)$ implies the degree of objects, p and q similar to attribute a. There are various feasible options and it is selected as follows [22].

$$R_a(p, q) = 1 - \frac{|a(p) - a(q)|}{|a_{\text{mx}} - a_{\text{mn}}|} \tag{9}$$

where σ_a^2 denotes the variance of attribute a, whereas a_{mx} and a_{mn} denote the maximum and minimum attribute values. The main aim of this approach is to quantify the lower and upper approximations of a decision class with NN of test object q, and offer the best clues for membership prediction of specific class. Specifically, when $(R \downarrow C)(q)$ is maximum, it shows every q's neighbours come under class C. In case of higher value for $(R \uparrow C)(q)$, then it implies that at least one neighbour should be in the class. The classification task is often computed for y as the initialization of τ to 0.

In order to execute a better classification task, the models result in decision class with optimal fuzzy lower and upper approximation memberships. The complexity of this model is $(|\mathcal{C}| \cdot (2|X|))$. It is implied as a zero order Takagi–Sugeno manager which is used as a rule, and the membership of the maximum test object. R_d shows fuzzy tolerance correlation for decision feature, d. Here, similar relationship is applied for conditional features. It is evident that with $\ell = \ell_M$ and $\mathcal{T} = \mathcal{T}_M$, condition $\tau_2 = 0$ is satisfied while $R(q, z) = 1$ for each neighbour z in N (total affinity of test object as well as NN). However, $R_d(z_1, z_2) = 0$ for each z_1, z_2 in N(total dissimilarity between the decision values of two neighbours).

As per relevant approximations of remarkable FR set theory, the methods are provided to resolve the presence of noise. This is because Sup and Inf are exploited and generalized with existential as well as universal quantifiers, correspondingly. On the other hand, it was not influenced by the choice of I, instead it was validated against crisp decisions with individual NN applied during classification task. It is advantageous in parameter election process which defines the classification decisions based on single object and it tends to produce maximum noisy data.

For this purpose, VQNN (Vaguely Quantified Nearest Neighbours) model, a FRSNN approach, is presented in which $R \downarrow C$ and $R \uparrow C$ were substituted by $R \downarrow^{Q_u} C$ and $R \uparrow^{Q_l} C$, correspondingly. Analogously, VQNN2 is an variant of FRSNN2, where $R \downarrow R_d z$ and $R \uparrow R_d z$ were exchanged by $R \downarrow^{Q_u} R_d z$ and $R \uparrow^{Q_l} R_d z$, respectively. For FRSNN, the application of K is not significant. In FRSNN, the effect has become minimal since $R(p, q)$ acquires smaller and p results in minimum influence on $(R \downarrow C)(q)$ and $(R \uparrow C)(q)$.

2.3 Optimal Rule Set Generation Using KH Algorithm

KH algorithm is applied to optimize the rule set generated by FRSNN model. KH is defined as a new meta-heuristic Swarm Intelligence (SI) optimization model. This model is used to resolve the optimization issues on the basis of herding nature outcomes attained from krill members [23]. Time-based position of a krill in 2D plane can be computed as three major events:

- Movement made by alternate krill members;
- Foraging behavior;
- External diffusion.

KH method has applied Lagrangian algorithm which is as follows.

$$\frac{dX_i}{dt} = N_i + F_i + D_i, \tag{10}$$

where N_i denotes the motion persuaded by alternate krill individuals; F_i represents the foraging action; and D_i shows external diffusion of ith krill individuals. The process involved in KH algorithm is demonstrated in Fig. 3.

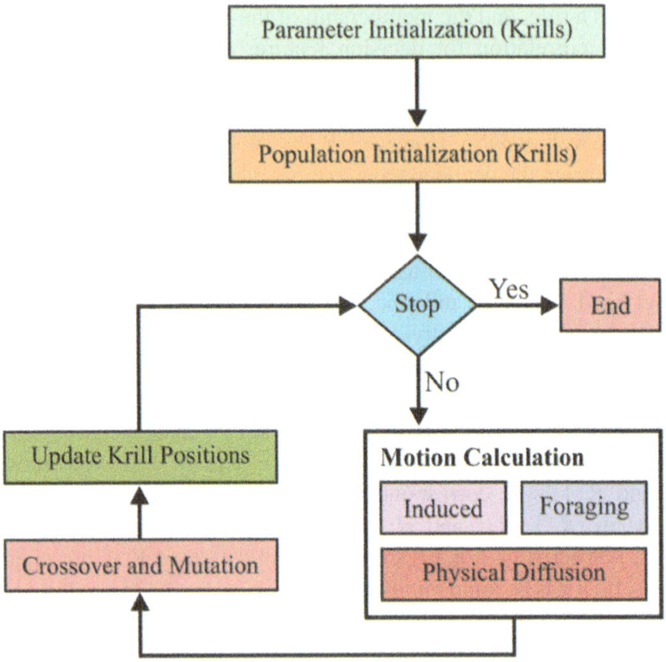

Fig. 3 Flowchart of KH algorithm

2.3.1 Motion Induced by Other Krill Individuals

In case of krill individual, the motion is encouraged by additional krill members which is implied as follows

$$\begin{aligned} N_i^{new} &= N^{\max}\alpha_i + \omega_n N_i^{old}, \\ \alpha_i &= \alpha_i^{local} + \alpha_i^{target}, \end{aligned} \tag{11}$$

where $N^{\max}$ denotes high induced speed, ω_n shows inertia weight of motion persuaded in [0, 1], N_i^{old} signifies final motion provided, α_i^{local} defines the local effect offered by neighbors, and α_i^{target} refers to the target dimension induced by optimal krill members. Consequently, the impact of neighbors could be estimated by,

$$\begin{aligned} \alpha_i^{local} &= \sum_{j=1}^{NN} \widehat{K}_{ij}\widehat{X}_{ij}, \\ \hat{X}_{ij} &= \frac{X_j - X_i}{\|X_j - X_i\| + \varepsilon}, \\ \hat{K}_{ij} &= \frac{K_j - K_i}{K^{worst} - K^{best}}, \end{aligned} \tag{12}$$

where K_i implies the fitness measure of ith krill individual. K^{best} and K^{worst} are superior and inferior fitness rates of krill individuals. K_j defines the fitness of jth$(j = 1, 2, \ldots, NN)$ neighbor. NN mimics the count of neighbours. X indicates the relevant position. The sensing distance of every krill individual is computed as given below:

$$d_{s,i} = \frac{1}{5N}\sum_{j=1}^{N}\|X_i - X_j\|, \tag{13}$$

where N corresponds to the value of krill members. The consequence of KH with optimal fitness on ith individual krill is assumed as given in this equation.

$$\partial_i^{target} = C^{best}\widehat{K}_{i,best}X_{i,best}, \tag{14}$$

where the measure of C^{best} is determined as:

$$C^{best} = 2\left(rand + \frac{I}{I_{\max}}\right) \tag{15}$$

2.3.2 Foraging Motion

This metric is computed with respect to two major parameters. Food location followed by the experience regarding first one are the two parameters involved in this metric. It can represent ith krill individual as given below.

$$F_i = V_f \beta_i + \omega_f F_i^{old},$$
$$\beta_i = \beta_i^{food} + \beta_i^{best}, \tag{16}$$

where y_f implies foraging speed, ω_f denotes the inertia weight of foraging action from 0 and 1, and F_i^{old} refers to consequent foraging action. β_i^{food} depicts the attraction of food and β_i^{best} shows the impact of better fitness of ith krill. Here, $y_f = 0.02$.

The virtual center of food concentration can be determined on the basis of fitness distribution of KH which is persuaded from the *center of mass* and is represented below.

$$X^{food} = \frac{\sum_{i=1}^{N}(1/K_i)X_i}{\sum_{i=1}^{N}(1/K_i)}. \tag{17}$$

Hence, the food concentration for ith KH is evaluated as given below.

$$\beta_i^{food} = C^{food}\widehat{K}_{i,food}\widehat{X}_{i,food}, \tag{18}$$

where C^{food} denotes the food coefficient using the given function.

$$C^{food} = 2\left(1 - \frac{I}{I_{\max}}\right). \tag{19}$$

The impact of optimal fitness of ith krill member is managed under the application of given formula:

$$\beta_i^{best} = \widehat{K}_{i,ibest}\widehat{X}_{i,ibest}, \tag{20}$$

where $\widehat{K}_{i,ibest}$ implies the previously-visited position of ith krill individual.

2.3.3 Physical Diffusion

The arbitrary diffusion of KH is assumed to be an additional random task. It is defined with respect to high diffusion speed and random directional vector as given herewith.

$$D_i = D^{\max}\delta, \tag{21}$$

where $D^{\max}$ denotes heavy diffusion speed, and δ defines random directional vector, and corresponding arrays are arbitrary measures in $[-1, 1]$. When the position is the best, the random motion becomes minimum. The impacts of motion encouraged by other krill individuals and foraging motion are reduced with improving iterations. Hence, an alternate Eq. (21) is substituted instead of Eq. (22). It gets linearly decreased with random speed and time and it functions according to geometrical annealing schedule.

$$D_i = D^{\max}\left(1 - \frac{I}{\mathrm{I}_{\max}}\right)\delta \tag{22}$$

2.3.4 Crossover

It is managed using crossover probability Cr, while mth component of $X_i(x_{i,m})$ is described herewith.

$$x_{i,m} = \begin{cases} x_{r,m} & rand_{i,m} < Cr \\ x_{i,m} & else, \end{cases}$$
$$Cr = 0.2\hat{K}_{i,best} \tag{23}$$

where $r \in \{1, 2, \ldots, i-1, i+1, \ldots, N\}$.

2.3.5 Mutation

It is handled by a mutation probability, Mu and an adaptive mutation approach is evaluated as given herewith.

$$x_{i,m} = \begin{cases} x_{gbest,m} + \mu(x_{p,m} - x_{q,m}) & rand_{i,m} < Mu \\ x_{i,m} & else, \end{cases}$$
$$Mu = 0.05\widehat{K}_{i,bets}, \tag{24}$$

where $p, q \in \{1, 2, \ldots, i-1, i+1, \ldots, K\}$, $\mu \in [0, 1]$.

2.3.6 Processes Involved in KH Algorithm

In general, the predetermined actions are primarily modified with position of krill members to attain optimal fitness. Foraging action, encouraged by alternate krill individuals, possesses two global and two local principles. Simultaneously, KH becomes an effective model. Under the application of diverse parameters, the motion at time and the position vector of a krill individual at interval t to $t+\Delta t$ is provided herewith.

$$X_i(t + \Delta t) = X_i(t) + \Delta t \frac{dX_i}{dt}. \tag{25}$$

It is evident that Δt is a vital constant and should be fixed appropriately on the basis of optimization problem. Δt can be attained easily with the following function.

$$\Delta t = C_t \sum_{j=1}^{NV} (UB_j - LB_j), \tag{26}$$

where NV denotes the variables and LB_j, UB_j denote the lower and upper bounds of jth variable ($j = 1, 2, \ldots, NV$) correspondingly. Here, C_t is a constant value in $[0, 2]$.

3 Performance Validation

To confirm the efficiency of the newly presented model, a sequence of processes was carried out using Python Programming language. The applied dataset and accomplished outcomes were examined under various scenarios.

3.1 Dataset

In order to verify the simulation outcome of the presented LSTM-EISC approach, a benchmark dataset i.e., SEMEVAL2018 Task-1Emotion Intensity Ordinal Classification dataset was employed in this study [24]. It is comprised of 4042 tweets with four instances like joy, fear, anger, and sadness. Among the tweets considered, 1074 tweets are categorized under Joy, 650 tweets belong to the class of fear, 991 tweets are grouped under anger and finally 555 tweets denote sadness. The dataset was classified into training and testing data in the proportion of 7.5:2.5. Figure 4 shows the data relevant to a dataset.

Fig. 4 Class distribution of the Sem-Eval 2018 dataset

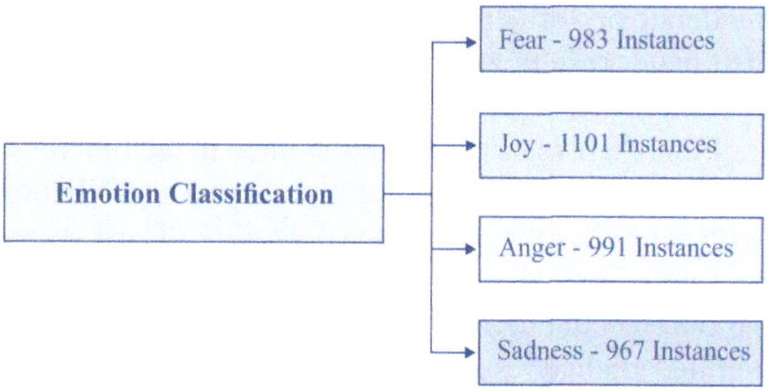

3.2 *Results*

Table 1 shows the results of the analysis of existing methods against the proposed FRSNN-KH model in terms of Precision, Recall, F-Measure and Accuracy. Figure 5 shows the performance results of FRSNN-KH algorithm in terms of precision. The figure states that the RF model produced an insignificant classifier outcome with a precision of 67.78%. The LR model attained 72.37% precision, somewhat better outcome to a certain extent. Besides, the SVM model achieved a

Table 1 Comparison of results attained by existing models and the proposed FRSNN-KH model under several measures

Methods	Precision	Recall	F-measure	Accuracy
Proposed FRSNN-KH	79.62	85.40	82.84	83.59
NB	75.63	74.47	75.05	75.19
RF	67.78	83.18	74.70	71.76
SVM	73.98	79.18	76.49	75.61
LR	72.37	78.30	75.21	74.15

Fig. 5 Precision analysis

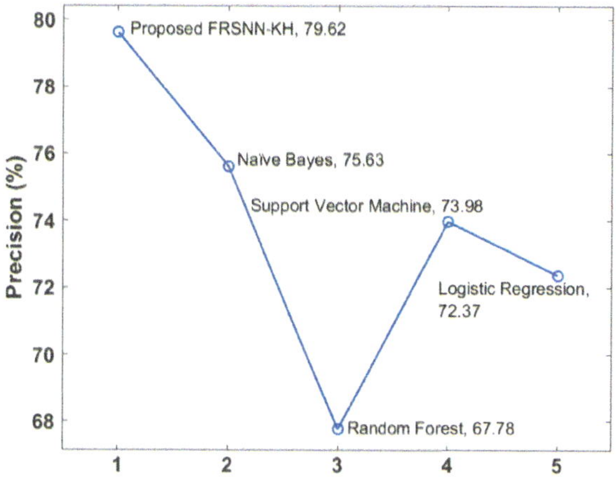

moderate performance with a precision of 73.98%. Moreover, the NB model exhibited better results with 75.63% precision. Furthermore, the FRSNN-KH algorithm accomplished supreme performance and offered the highest precision of 79.62%.

Figure 6 reveals the results attained in the functioning of FRSNN-KH model with respect to recall. The figure reveals that the NB model is an ineffective classifier with a recall value outcome of 74.47%. The LR model exhibited moderate result to definite extent and reached 78.30% recall. On the other end, the SVM method produced better performance with a recall of 79.18%. Additionally, the RF technique implied gradual results with a recall of 83.18%. However, the presented FRSNN-KH approach accomplished the best performance with an optimal recall of 85.40%.

Figure 7 refers to the performance results of the presented FRSNN-KH model by means of F-measure. From the figure, it is clear that the RF model produced inferior classifier outcome i.e., 74.70% F-measure. Followed by, the NB model produced considerable results and achieved F-measure of 75.05%. On the other hand, the LR

Fig. 6 Recall analysis

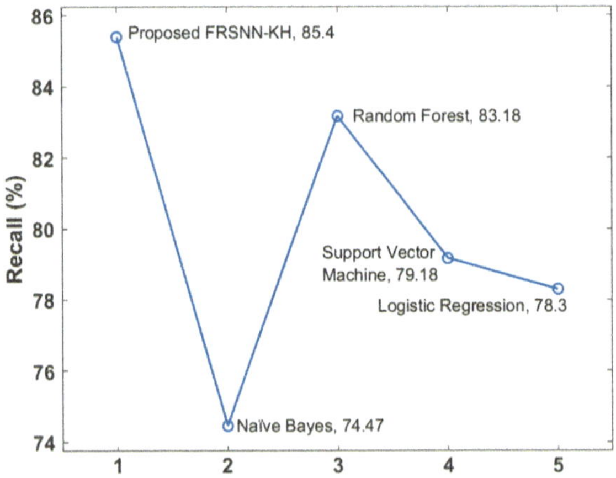

Fig. 7 F-measure analysis

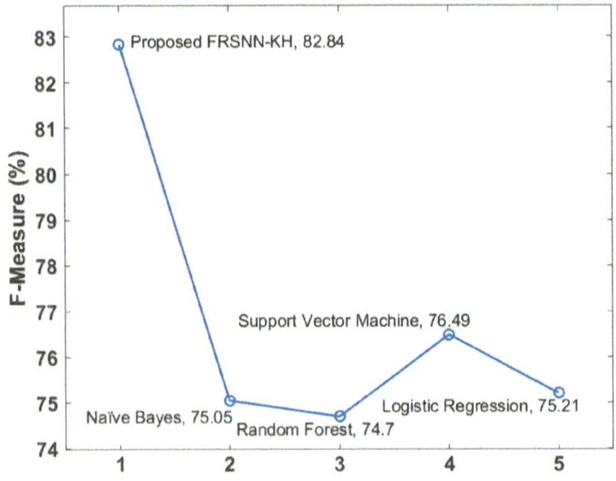

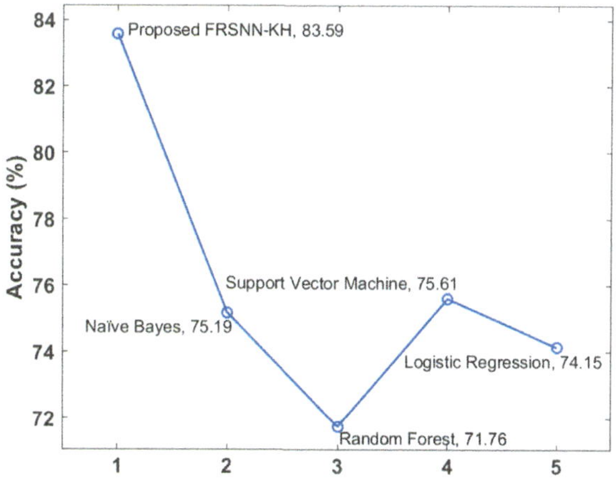

Fig. 8 Accuracy analysis

model accomplished better performance with an F-measure of 75.21%. Additionally, the SVM model implied moderate results with 76.49% F-measure. However, the presented FRSNN-KH technique achieved an optimized performance and provided the best F-measure of 82.84%.

Figure 8 depicts the performance results of the presented FRSNN-KH method in terms of accuracy. The figure refers that the RF model is an ineffective classifier as it achieved 71.76% accuracy. Additionally, the LR model depicted a manageable outcome and achieved an accuracy of 74.15%. Followed by, the NB model implied a gradual performance increment with 75.19% accuracy. The SVM model produced an accuracy of 75.61%, appreciable than previous models. However, the FRSNN-KH model accomplished superior performance and high accuracy of 83.59%.

From the above-mentioned tables and figures, it is inferred that the proposed FRSNN-KH algorithm is an excellent tool for classification of sentiments since it outperformed other models in terms of precision (79.62%), recall (85.40%), F-measure (82.84%) and accuracy (83.59%).

4 Conclusion

The current research work developed an efficient sentiment and emotion classification model named FRSNN-KH algorithm. The proposed FRSNN-KH algorithm has three steps namely preprocessing, feature extraction and classification. The input data was first pre-processed and then BoW method was applied as a feature extractor to extract the useful features from pre-processed tweet. Then, FRSNN-KH-based classification was carried out to segregate the instances under different class labels. Soft computing-based KH algorithm was applied in the optimization of number of rules generated by FRSNN model. This further helped in improving classification performance. The performance of the FRSNN-KH algorithm was validated using the benchmark dataset and the results were assessed under several aspects. The results inferred that the proposed algorithm is an excellent tool for classification of sentiments. In future we will try to involve recurrent model driven emotion and sentiment classification model with gated inputs and outputs.

References

1. Kumar, A., Sebastian, T.M.: Sentiment analysis on twitter. Int J Comput Sci Issues. **9**(4), 372–378 (2012)
2. Dave, K., Lawrence. S., Pennock, D.M.: Mining the peanut gallery: opinion extraction and semantic classification of product reviews. In: Proceedings of the 12th international conference on World Wide Web. Budapest, Hungary (2003)
3. Reilly, T.: Web 2.0 Compact Definition: Trying Again. Sebastopol, CA: O'Reilly Media. http://radar.oreilly.com/2006/12/web-20-compactdefinition-tryi.html (2017). Accessed 24 April 2017
4. Alfouzan, H.I.: Introduction to SMAC-social mobile analytics and cloud. Int J Sci Eng Res. **6**, 128–130 (2015)
5. Shankar, K., WahabSait, A.R., Gupta, D., Lakshmanaprabu, S.K., Khanna, A., Pandey, H.M.: Automated detection and classification of fundus diabetic retinopathy images using synergic deep learning model. Pattern Recogn. Lett. **133**, 210–216 (2020)
6. Pustokhina, I.V., Pustokhin, D.A., Gupta, D., Khanna, A., Shankar, K., Nguyen, G.N.: An effective training scheme for deep neural network in edge computing enabled Internet of Medical Things (IoMT) systems. IEEE Access **8**(1), 107112–107123 (2020)
7. Raj, J.S., Jeya Shobana, S., Pustokhina, I.V., Pustokhin, D.A., Gupta, D., Shankar, K.: Optimal feature selection based medical image classification using deep learning model in internet of medical things. IEEE Access **8**(1), 58006–58017 (2020)
8. Shankar, K., Lakshmanaprabu, S.K., Khanna, A., Tanwar, S., Rodrigues, Joel J.P.C., Roy, N.R.: Alzheimer detection using group grey wolf optimization based features with convolutional classifier. Comput. Electr. Eng. **77**, 230–243 (2019)
9. Al-smadi, M., Qawasmeh, O., Al-ayyoub, M., Jararweh, Y., Gupta, B.: Deep recurrent neural network vs. support vector machine for aspect-based sentiment analysis of Arabic hotels' reviews. J. Comput. Sci., 386–393 (2017). https://doi.org/10.1016/j.jocs.2017.11.006
10. Al-Smadi, M., Talafha, B., Al-Ayyoub, M., Jararweh, Y.: Using long short-term memory deep neural networks for aspect-based sentiment analysis of Arabic reviews. Int. J. Mach. Learn. Cybern., 1–13 (2018). https://doi.org/10.1007/s13042-018-0799-4
11. Al-ayyoub, M., Al-smadi, M., Al-ayyoub, M., Jararweh, Y., Qawasmeh, O.: Enhancing aspect-based sentiment analysis of Arabic hotel' reviews using morphological, syntactic and semantic features. Inf. Process. Manage. 308–319 (2018). https://doi.org/10.1016/j.ipm.2018.01.006
12. García-Pablos, A., Cuadros, M., Rigau, G.: W2VLDA: almost unsupervised system for aspect based sentiment analysis. Expert Syst. Appl. **91**, 127–137 (2018)
13. Dragoni, M., Federici, M., Rexha, A.: An unsupervised aspect extraction strategy for monitoring real-time reviews stream. Inf. Process. Manage. 1103–1118 (2018)
14. Rathan, M., Hulipalled, V.R., Venugopal, K.R., Patnaik, L.M.: Consumer insight mining: aspect based twitter opinion mining of mobile phone reviews. Appl. Soft Comput. J. **68**, 765–773 (2018). https://doi.org/10.1016/j.asoc.2017.07.056
15. Cambria, E., Poria, S., Hazarika, D., Kwok, K.: SenticNet 5: discovering conceptual primitives for sentiment analysis by means of context embeddings. In: Thirty Second AAAI Conference on Artificial Intelligence, pp. 1795–1802 (2018). https://www.aaai.org/ocs/index.php/AAAI/AAAI18/paper/view/16839. Accessed 26 Oct 2019

16. Xianghua, F., Guo, L., Yanyan, G., Zhiqiang, W.: Multi-aspect sentiment analysis for Chinese online social reviews based on topic modeling and HowNet lexicon. Knowl. Based Syst., **37**, 186–195 (2013). https://doi.org/10.1016/j.knosys.2012.08.003
17. Chen, Z., Mukherjee, A., Liu, B.: Aspect extraction with automated prior knowledge learning. In: Proceedings of the 52nd Annual Meeting of the Association for Computational Linguistics (Volume 1: Long Papers), pp. 347–358 (2014)
18. Poria, S., Cambria, E., Ku, L.-W., Gui, C., Gelbukh, A.: A rule-based approach to aspect extraction from product reviews. In: Proceedings of the Second Workshop on Natural Language Processing for Social Media (SocialNLP), pp. 28–37 (2014)
19. Pham, D., Le, A.: Data & knowledge engineering learning multiple layers of knowledge representation for aspect based sentiment analysis. Data Knowl. Eng. **114**, 26–39 (2018). https://doi.org/10.1016/j.datak.2017.06.001
20. Liu, Q., Gao, Z., Liu, B., Zhang, Y.: Automated rule selection for aspect extraction in opinion mining. In: IJCAI, pp. 1291–1297 (2015)
21. Ma, Y., Peng, H., Cambria, E.: Targeted aspect-based sentiment analysis via embedding common sense knowledge into an attentive LSTM, In: Thirty Second AAAI Conference on Artificial Intelligence, pp. 5876–5883 (2018). https://www.aaai.org/ocs/index.php/AAAI/AAAI18/paper/view/16541. Accessed 14 Sept 2019
22. Jensen, R., Cornelis, C.: Fuzzy-rough nearest neighbour classification and prediction. Theoret. Comput. Sci. **412**(42), 5871–5884 (2011)
23. Gandomi, A.H., Alavi, A.H.: Krill herd: a new bio-inspired optimization algorithm. Commun. Nonlinear Sci. Numer. Simul. **17**(12), 4831–4845 (2012)
24. Source: https://competitions.codalab.org/competitions/17751

Fuzzy Wavelet Neural Network with Social Spider Optimization Algorithm for Pattern Recognition in Medical Domain

Sujatha Krishanmoorthy, S. Selvakumar, Veeraraghavan Vishnu Priya, Surapaneni Krishna Mohan, and K. Sabithabanu

Abstract Providing accurate and accessible diagnoses is an important challenge for global healthcare system. The recently developed artificial intelligence (AI) and machine learning (ML) have emerged as powerful tools for solving medical data classification problems. The current research article presents a new Fuzzy Wavelet Neural Network (FWNN) with Social Spider Optimization (SSO) algorithm to recognize the medical data. The aim of FWNN is to utilize both WNN and Adaptive Fuzzy Neuro Inference System (ANFIS) to increase the convergence rate and achieve better approximation abilities. But a number of parameters should be adjusted in due course. In addition, the fuzzy rules enable the technique to tackle uncertainty, whereas the wavelets contribute to enhancing the accuracy of the input output map. SSO algorithm is applied to optimize the architectural parameters of FWNN model such as learning parameters, weights, dilation, and translation. The application of SSO algorithm helps in determining the hyperparameters of FWNN model during medical diagnosis. This is the novelty of current work and it assists in improving the classification performance. The SSO-FWNN model was experimentally validated using three medical benchmark datasets namely, breast cancer, heart disease, and hepatitis. The outcome

S. Krishanmoorthy (✉)
Department of Computer Science, Wenzhou Kean University, Wenzhou, Zhejiang Province, China
e-mail: sujatha@wku.edu.cn

S. Selvakumar
Department of Information and Technology, Dhanalakshmi Srinivasan Engineering College, Perambalur, India

V. Vishnu Priya
Department of Biochemistry, Saveetha Dental College, Saveetha Institute of Medical and Technical Sciences, Chennai, India

S. Krishna Mohan
Department of Biochemistry, Panimalar Medical College Hospital & Research Institute, Varadharajapuram, Poonamallee, Chennai, India

K. Sabithabanu
Department of Electronics and Communication Engineering, Mohamed Sathak A. J. College of Engineering, Chennai, India

D. Gupta et al. (eds.), *Soft Computing for Data Analytics, Classification Model, and Control*, Studies in Fuzziness and Soft Computing 413,
https://doi.org/10.1007/978-3-030-92026-5_2

of simulations indicates that the SSO-FWNN model achieved maximum outcome on all the applied datasets over existing methods in a significant way.

Keywords Classification · Fuzzy logic · Pattern recognition · Soft computing · Wavelet neural network

1 Introduction

In recent days, knowledge based expert systems have been developed using artificial intelligence techniques which aim to assist the diagnostic decision-making ability of human experts in a short timespan [1]. Several soft computing algorithms have evolved in recent years, e.g., Genetic Algorithms (GAs) evolved on the basis of Darwinian Evolutionary Algorithm (EA), Ant Colony Optimization (ACO) which resembles the character of ants foraging for food, Particle Swarm Optimization (PSO) implies the nature of birds flocking together, Simulated Annealing (SA) under the application of thermodynamic impacts, Artificial Bee Colony algorithm (ABC), an optimization model that works according to the smart foraging behavior of honey bee swarm, Bat Algorithm (BA) evolved with the echolocation behaviour of micro bats [2], and so on. Likewise, Gravitational Search Algorithm (GSA) is defined as a heuristic optimization algorithm that functions on the basis of law of gravity and mass d. It depends upon Newtonian gravity which is defined that "each particle in the universe attract every other particle with a force that is directly proportional to the product of their mass and inverse proportion of the square of the distance among them" [3].

Earlier studies have also focused on the design of GSA in different application areas [3–5]. Further, a fuzzy GSA (FGSA) has also been presented earlier as the best model for multi-machine Power System Stabilizers (PSSs) [4]. FGSA-driven PSS method has been verified for two multi-machine systems such as 3-machine 9-bus system and 10-machine 39-bus system. Hence, Fuzzy Logic (FL) is applied to improve the final part of the system and explore the exact solution within the given time interval, especially for huge issues. In the newly-developed GSA, fuzzy-relied method and fitness sharing have been used in decision-making process i.e., during the selection of optimal solution from Pareto front. But it presents a novel FGSA to process the dynamic modifications at alpha parameter. It also provides gravitation as well as acceleration measures to all the agents who tend to enhance the efficiency of the system, as cited in the literature [5]. It is mainly applied in designing the best Modular Neural Networks (MNNs) for pattern analysis that hasn't been utilized in classical methods.

In the last few decades, medical diagnosis has developed rapidly, since it is a non-invasive mechanism in clinical diagnoses. Hence, FGSA is exploited to establish a better structure for MNN types which aims at identifying the best amount of nodes, modules, and layers of NN for echocardiogram (ECG) realization. At this point, ECG database is applied with images retrieved from both normal and abnormal patients.

The major complication that hits the medical sector is the presence of abundant number of noisy ultrasound images. The consequent images get ruined due to the existence of noise which further degrades the edges of cardiac images [6]. In order to remove the noise, image processing is executed. After this process, the image achieves the best accuracy for pattern recognition. On the other end, both noise and alternate aspects create an impact on the results of ultrasound image analysis. Therefore, disease diagnosis becomes a challenging issue in cardiology. Moreover, 2D echo videos have inferior image quality due to minimum contrast, speckle noise, as well as signal limitations which altogether result in image interpretation complexities [7].

Initially, Shahram et al. [8] presented a constellation-of-parts dependent technique called automated cardiac view recognition model. It was applied in generic heart chamber detection device [9] which helps in placing the heart chambers at appropriate place. Further, it implied a spatial organization under the application of Markov Random Field (MRF)—relied on relational graph. Also, the researchers conducted a last classification task in a sample image through Support Vector Machine (SVM) on the basis of MRF system results. Unfortunately, this model experienced sensitivity issues in chamber detection process. The noise present in ECG images, at the time of depiction, restricted the robustness in actual image conversions. Dwivedi et al. [10] deployed a model for automated exploration of heart ECG video reviews. Here, the heart shape is implied through a constellation of chambers under distinct perceptions. The statistical difference between the constellation parts and corresponding spatial associations was developed with the application of MRF techniques.

Additionally, discriminative model is applied for view examination that combines the evaluations of sample images by view models. Banning et al. [11] and Sugioka [12] presented an approach to predict cardiac structures using patterns in which active contours (snakes) are applied in ECG images. Alternatively, Comaniciu et al. [13] implied a model to observe cardiac boundaries in ECG images using different extracts of the image acquired. Wu et al. [14] projected a model for ECG view classification by utilizing low-level features. It provides a view classification model for 2D heart ultrasound data. Furthermore, it applied lower level image features for training frame-level classification i.e., contrary-based modules. This does not require an extra pixel-level classification of heart images. With the application of kernel-relied classification, the methods classify the images from any stage of heartbeat cycle and produce data from consequent frames with no need for a re-training approach.

Dowlatshahi and Nezamabadi-pour [15] applied a technique named Grouping GSA (GGSA) to perform data clustering. Being a newly presented model, it differs from standard GSA with two vital factors. Initially, a particular encoding approach termed 'grouping encoding' was mainly applied in the development of related structures of clustering issues which are the portions of solutions. Then, the encoding for special GSA upgrading equations was applied for solutions with 'grouping encoding'. The function of the presented method was calculated using different reputed datasets from popular UCI Machine Learning (ML) Repository. Through image pre-processing, the size of the image was reduced from 200×125 to 80×80 pixels.bmp. Here, the Region of Interest (ROI) is assumed by eliminating major amount of noisy portions in the image.

The current research article develops an effective medical data classification model named Fuzzy Wavelet Neural Network (FWNN) with Social Spider Optimization (SSO) algorithm. This SSO-FWNN model is applied for the optimization of hyperparameters of FWNN model. The model assists in improving the classification performance. The SSO-FWNN model was experimentally validated using three medical benchmark datasets namely, breast cancer, heart disease, and hepatitis. The outcome of simulations indicates that the SSO-FWNN model achieved maximum outcome on all the applied datasets over existing approaches in a considerable manner.

2 The Proposed SSO-FWNN Model

Figure 1 illustrates the work processes of SSO-FWNN system. The proposed SSO-FWNN technique consists of two main procedures namely, FWNN-based classification, and SSO algorithm for hyperparameter optimization. These processes are clearly discussed in the following subsections.

2.1 FWNN Model

FWNN model is composed of seven layers as briefed herewith.

Layer 1: No processing in this layer, thereby every node directly transmits the input values to second layer.

Layer 2: The measures of input parameters are fuzzified and every node implies a single fuzzy set. Hence, the resultant of layer 2 shows the membership values of input

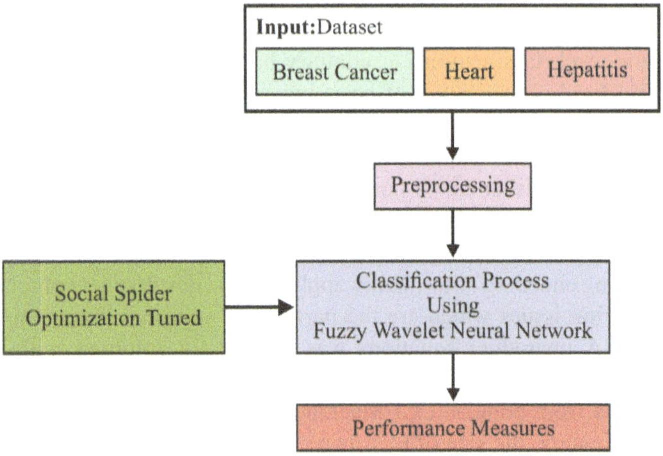

Fig. 1 Overall process of SSO-FWNN model

variable to the respective fuzzy set, according to the decided Membership Function (MF) [16].

Layer 3: This layer accepts the MF of inputs, acquired from existing layer, to develop the antecedent portion of every fuzzy rule, in which all the nodes define one fuzzy rule. The simulation outcome of this layer is computed using the Eq. (1) given below

$$\mu_i = (x) = \prod_{j=1}^{q} A_j^i(x_j) \tag{1}$$

where Π is a product task and A_j^i is treated as an MF and is applied to determine the membership degree of input parameters. Here, a Gaussian MF is applied and is shown in Eq. (2).

$$A_j^i(x_j) = \exp\left[-\left(\frac{x_j - c_j^i}{\sigma_j^i}\right)^2\right] \tag{2}$$

In Eq. (2), c_j^i and σ_j^i signify center as well as width of an MF respectively.

Layer 4: The nodes present in layer 4 are named after 'consequent nodes' as they correspond to the final portion of fuzzy rules. Every node is comprised of wavelet function (ψ_{ij}). The wavelet function is defined as a waveform with minimum time interval and the mean value is 0. Further, it has dilation and translation parameters. In FWNN, fuzzy rules are given by:

$$R^i : if\, x1\, is\, A_1^i, and\, x2\, is\, A_2^i \ldots and\, xq\, is\, A_q^i\, then\, y_i = w_i \sum_{j=1}^{q} \psi_{ij}(x_j) \tag{3}$$

where R^i is the ith rule. Here, $(1 \le i \le c)$ c defines the count of a rule. x_j denotes the jth input, and $(1 \le j \le q)$, q represents the count of input parameter. Further, y_i implies the ith output from WNN model. In WNN, the final outcome can be determined by applying Eq. (4).

$$y_i = w_i \sum_{j=1}^{q} \psi_{ij}(x_j) \tag{4}$$

where w_i showcases the weight coefficients and ψ_{ij} demonstrates a collection of wavelet functions, named as wavelet family which is computationally described by Eq. (5).

$$\psi_{i_j}(x_j) = \psi\left(\frac{x_j - b_{ij}}{a_{ij}}\right), \ a_{i_j} \neq 0 \tag{5}$$

where a_{ij} and b_{ij} are dilation and translation variables of Wavelet Function (WF), correspondingly, Here, ψ implies the mother wavelet estimated under the application of Mexican hat WF as expressed through Eq. (6).

$$\psi(x) = \frac{1}{\sqrt{|a|}}(1 - 2x^2) \tag{6}$$

Layer 5: It is a combination of layer 3 and layer 4. The resultant value is computed by maximizing the simulation outcome of 3rd layer with the results of 4th layer by applying Eq. (7).

$$L_5(i) = \mu_i^* y_i \tag{7}$$

In which $1 \leq i \leq c$ and c signifies the amounts of fuzzy rules and wavelet functions.

Layer 6: The outcome of these layers is comprised of 2 portions. Initially, (L_6^a) is calculated by adding the results of 5th layer as shown in Eq. (8)

$$L_6^a = \sum_{i=1}^{c} \mu_i^* y_i \tag{8}$$

Secondly, (L_6^b) is determined by adding the results from 3rd layer and is expressed by Eq. (9)

$$L_6^b = \sum_{i=1}^{c} \mu_i \tag{9}$$

Layer 7: It is represented as a defuzzifier whereas the entire results of FWNN are computed using Eq. (10)

$$\mathcal{U} = \frac{\sum_{i=1}^{c} \mu y_i}{\sum_{i=1}^{c} \mu_i} \tag{10}$$

2.2 SSO Algorithm

It is a novel optimization model that has been developed based on the evolution of trained strategy of spiders' social nature [16]. The solution space, present in SS, is a community network in which the spider interacts. The main features involved in this approach are as follows.

- In community network, a solution within a solution space illustrates the position of the spider.
- To decide the optimal fitness solution rate, each spider is positioned based on its weight.
- Female and male are 2 types of searching agents developed by this model. All the individuals are operated at diverse points, according to the genre by their social behavior obtained in the community.

The major features of SS communities are as follows: it is composed of massive females. Overall, the female N_f is around 60–95% by the given statement. Regarding N, the male N_m is computed as N_f complement. The whole population S is classified to M and F, in which M denotes the male spider and F implies the female spider. Figure 2 shows the flowchart of SSO technique [17].

$$F = \{f_1, f_2, \ldots, f_{N_f}\} \tag{11}$$

$$M = \{m_1, m_2, \ldots, m_{N_m}\} \tag{12}$$

where $S = F \cup M = \{s_1, s_2, \ldots, s_N\}$.

2.2.1 Calculation of Fitness

In population S, the superiority of the solutions are depicted by w_x weight received via individual (spider) x. Weight processing can be formulated as given herewith.

$$w_x = \frac{J(s_x) - worst_s}{best_s - worst_s} \tag{13}$$

where fitness value is depicted by $J(s_x)$ under the examination of spider position s_x about a function, J. The values of $best_s$ and $worst_s$ are maximization problems which are represented as follows.

$$best_s = \max(J(s_z)),\ k \in \{1, 2, \ldots, N\} \tag{14}$$

$$worst_s = \min(J(s_z)),\ k \in \{1, 2, \ldots, N\} \tag{15}$$

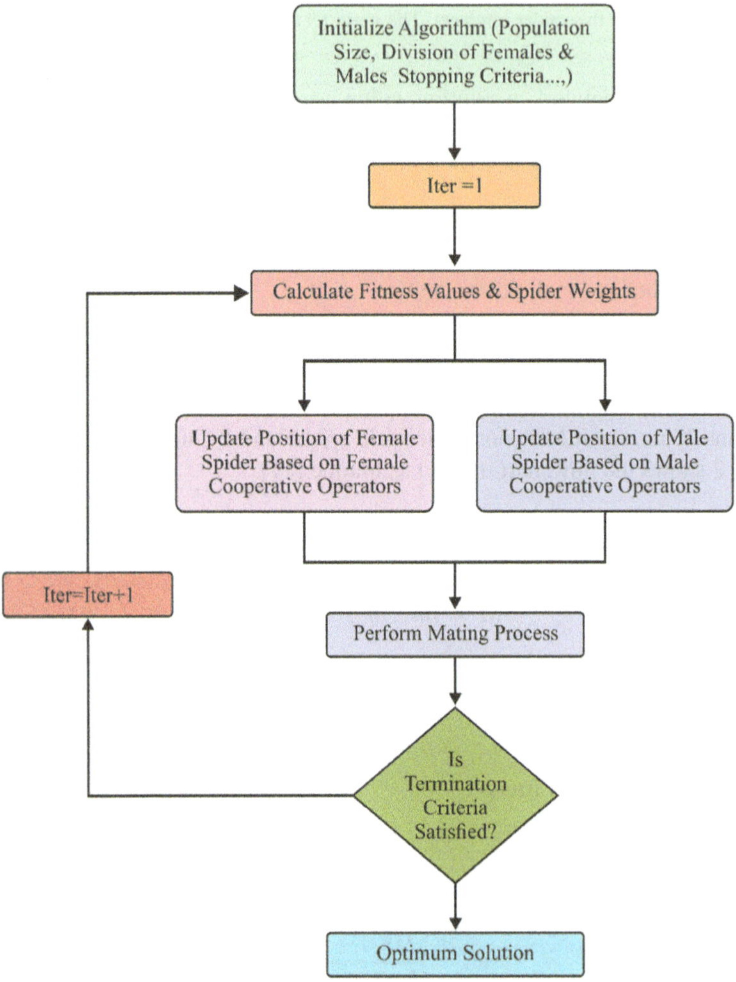

Fig. 2 Flowchart of SSO algorithm

2.2.2 Modeling of Vibrations Through Community Network

In order to transfer the data from colony participants, the function of community network is applied. It is complex in each individual that embeds the data by small vibrations. The vibrations are based on distance as well as weight of the spiders. The vibrations, employed by the individual $\boldsymbol{x}$ when data is transferred via individual $\boldsymbol{y}$, are developed as follows

$$\boldsymbol{Vib}_{x,y} = \boldsymbol{w}_y * e^{-d^2_{x,y}} \tag{16}$$

Assume $\boldsymbol{d}_{x,y}$ is the Euclidean distance amongst two spiders, $\boldsymbol{y}$ and $\boldsymbol{x}$. There are 3 types of vibration that exist in SS method.

- $\boldsymbol{Vib}_{x,c}$ vibration in which c denotes the member nearby $\boldsymbol{x}$ that has improved its weight than $\boldsymbol{x}(\boldsymbol{w}_c > \boldsymbol{w}_x)$
- $\boldsymbol{Vib}_{x,b}$ vibration in which b represents the individual with optimal weight of overall population S.
- $\boldsymbol{Vib}_{x,f}$ vibration in which f denotes the female individual identical to $\boldsymbol{x}$.

2.2.3 Initialization of Population

By initiating the set S, the SS method begins with locations of N spiders. For n-dimensional vector, each position $\boldsymbol{m}_x$ or $\boldsymbol{f}_x$ with rates has to be improvised. Upon the rates, $\boldsymbol{p}^{high}$ and $\boldsymbol{p}^{low}$, it is evenly distributed as mentioned.

2.2.4 Cooperative Operators

Cooperative Operator for Female Spiders

A new task is carried out to estimate the cooperative nature of female spiders. For all iterations, the task ensures the modification of $\boldsymbol{x}$ position of the female spider. It may be either repulsive or attractive, which can be computed using three elements.

- The basic element considers the change based on nearby $\boldsymbol{x}$ that is comprised of heavy weight and provides vibration, $\boldsymbol{Vib}_{x,c}$;
- In population S, the future component applies the alteration using optimal individual that results in vibration, $\boldsymbol{Vib}_{x,b}$;
- The upcoming results are combined with random movement.

Consequently, three members are considered as single movement, and "–" symbol is applied for repulsion and "+" for attraction. The modification of the position is calculated as follows.

$$f_x^{z+1} = f_x^z \pm \boldsymbol{movement} \tag{17}$$

where z denotes the iteration count.

Cooperative Operator for Male Spiders

The cooperative nature of the male spider can be followed by the segmentation of spiders under two colonies: dominant (D) and non-dominant (ND). In case of position, this classification is developed with male individuals. Dominants have maximum weight than the average; otherwise, it is named as 'non-dominant'. The 2 motions are provided for D males:

- The trend of attraction to closer female that develops a vibration, $\boldsymbol{Vib_{x,f}}$,
- An arbitrary movement

These two objectives are considered as 1 and alteration in a D male can be determined as follows.

$$\boldsymbol{m_x^{z+1} = m_x^z + D\_movement} \tag{18}$$

For weighted male spiders' value, a single motion of attraction is defined for ND males which are computed as follows.

$$\boldsymbol{m_x^{z+1} = m_x^z + ND\_movement} \tag{19}$$

2.2.5 Mating Operator

Mating in a spider colony is conducted between D females and males. With specific mating range, $\boldsymbol{r}$, when a D male $\boldsymbol{m_g}$ identifies a female spider set $\boldsymbol{E^g}$, then it develops new offspring $\boldsymbol{S_{new}}$ by mating. From a set $\boldsymbol{T^g}$, a novel offspring are generated by combining $\boldsymbol{m_g}$ and $\boldsymbol{E^g}$. The mating task is removed, when set $\boldsymbol{T^g}$ is null. The fresh offspring is influenced by weight of every spider added during mating step from set $\boldsymbol{T^g}$. Under the application of roulette-wheel selection, the influence probability $\boldsymbol{P_{s_x}}$ is assigned as

$$\boldsymbol{P_{s_x}} = \frac{\boldsymbol{w_x}}{\sum_{y \in T^g} \boldsymbol{w_y}} \tag{20}$$

where $\boldsymbol{s_x} \in \boldsymbol{T^g}$. In solution space, a spider is defined as a solution. The variables are selected using roulette model. For example, let $\boldsymbol{s_{new}} = \{\boldsymbol{v_1}, \boldsymbol{v_2}, \ldots, \boldsymbol{v_n}\}$ be a new spider, each variable $\boldsymbol{v_x}$ is selected by roulette-wheel selection approach. Using the worst spider $\boldsymbol{s_{worst}}$, a new spider $\boldsymbol{s_{new}}$ is related to weights, where $\boldsymbol{w_{worst}} = \mathbf{min}_{l \in \{1,2,\ldots,N\}}(\boldsymbol{w_l})$. While relating the new spider $\boldsymbol{s_{new}}$ is best, it is converted by $\boldsymbol{s_{new}}$. When the new spider is removed, then no modifications are carried out. Both genre and index of the novel spider are obtained from previous ones.

2.3 Hybridization of Parameter Optimization of FWNN Model

In order to optimize parameters of FWNN, SSO algorithm is employed and the processes involved are listed below.

(1) Assume the values of chosen dataset as input of FWNN model.
(2) Initialization of every spider (weight, dilation parameter, translation parameter) takes place in a random way that lies in the range of [−1, 1]. Afterward, the initialization of SSO algorithm occurs namely, maximum iteration number, population size, and dimension.
(3) Consider Cycle = 1.
(4) Determine the fitness function of SS.
(5) Updates the weights, dilations, and translation parameters by SSO algorithm.
(6) Pick the optimal SS with optimal fitness values as a global solution, indicating the optimal value of translation, weight, and dilation variables.
(7) Assume Cycle = Cycle + 1.
(8) Reiterate steps (3) to (6) till the predefined stopping criteria is satisfied.

3 Performance Validation

A series of research was carried out to verify the goodness of SSO-FWNN technique using three benchmark datasets. The details of the dataset are given in Table 1. The first breast cancer datasets have a total of 683 instances, 9 features, and 2 class labels. The second heart disease dataset has a total of 297 instances, 13 features, and 2 class labels. Lastly, the hepatitis dataset has a total of 80 instances, 19 features, and 2 class labels.

Table 2 and Figs. 3, 4 show the results of SSO-FWNN model upon the applied breast cancer dataset. When measuring the performance of SSO-FWNN model based on specificity and sensitivity, the values should be high to infer its performance as excellent. Among all the models considered, the DT model offered the least classifier outcome with 96.05% sensitivity and 91.76% specificity. In line with this, LR technique achieved a slightly effective outcome over DT model with 97.37% of sensitivity and 95.02% of specificity. Afterward, the RBFNetwork model showcased moderate results with a sensitivity of 98.20% and specificity of 91.73% correspondingly. Likewise, the RF model achieved a near optimal classifier outcome with 98.23%

Table 1 Dataset descriptions

Dataset	No. of instances	No. of features	No. of classes
Breast cancer	683	9	2
Heart	297	13	2
Hepatitis	80	19	2

Table 2 Results of SSO-FWNN model against existing methods for breast cancer dataset

Methods	Sensitivity	Specificity	Accuracy	Kappa
SSO-FWNN	98.56	96.43	97.54	94.98
RBFNetwork	98.20	91.73	95.85	90.95
LR	97.37	95.02	96.56	92.40
DT	96.05	91.76	94.56	87.99
RF	98.23	94.33	96.85	93.07

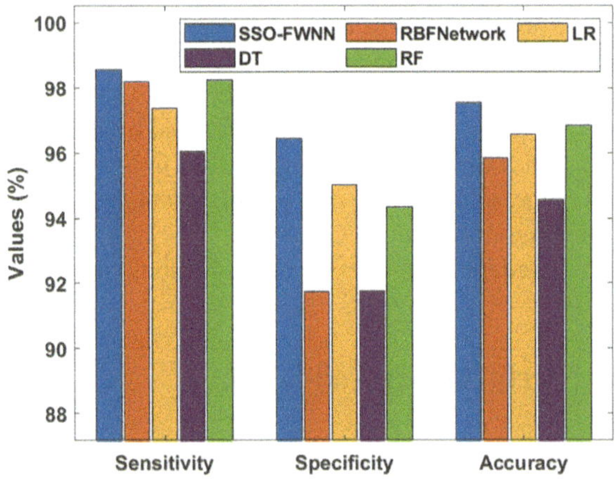

Fig. 3 Breast cancer analysis of proposed SSO-FWNN method

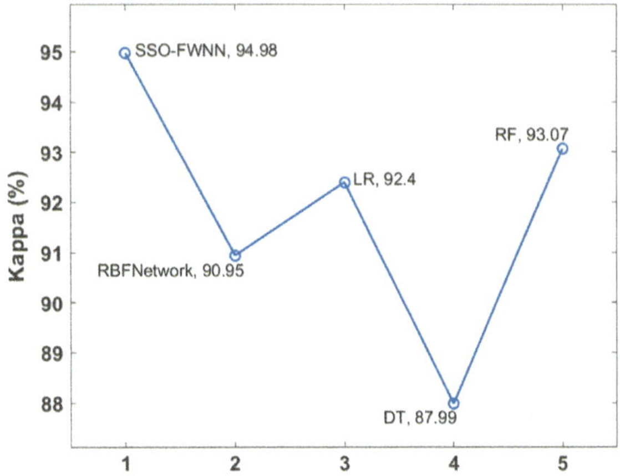

Fig. 4 Kappa analysis of proposed SSO-FWNN method

of sensitivity and 94.33% of specificity correspondingly. But the proposed SSO-FWNN model accomplished effective results over previous models with 98.56% of sensitivity and 96.43% of specificity correspondingly.

When determining the function of SSO-FWNN method with respect to accuracy and kappa, the values should be maximum to infer its performance as excellent. In comparison with other models, it shows that the DT model provided a minimum classifier result with an accuracy of 94.56% and kappa of 87.99%. In line with this, the RBFNetwork system achieved moderate results than DT model in terms of accuracy (95.85%) and kappa (90.95%). Then, the LR model displayed better results with an accuracy of 96.56% and kappa of 92.40%. As well as, the RF technique attained a close optimal classifier outcome with 96.85% of accuracy and kappa of 93.07% correspondingly. However, the presented SSO-FWNN model achieved active results in comparison with existing approaches with an accuracy of 97.54% and kappa of 94.98% respectively.

Table 3 and Figs. 5, 6 portray the results of SSO-FWNN model on the applied heart disease dataset. When computing the performance of SSO-FWNN method with respect to sensitivity and specificity, the performance can be inferred as effective only when high values are achieved. The overall comparison shows that the RT model afforded the least classifier outcome with 74.41% of sensitivity and 71.11% of specificity correspondingly. Similarly, the J48 model exhibited gradual outcome over RT model with a sensitivity of 78.94% and specificity of 78.03%. Next, the

Table 3 Results of SSO-FWNN method against existing methods for heart disease dataset

Methods	Sensitivity	Specificity	Accuracy	Kappa
SSO-FWNN	86.87	84.85	85.32	78.90
J48	78.94	78.03	78.55	56.64
RT	74.41	71.11	72.94	45.38
RBFNetwork	85.19	81.56	83.49	66.81
NBTree	80.24	77.94	79.21	58.06
RF	81.98	82.44	82.18	63.95

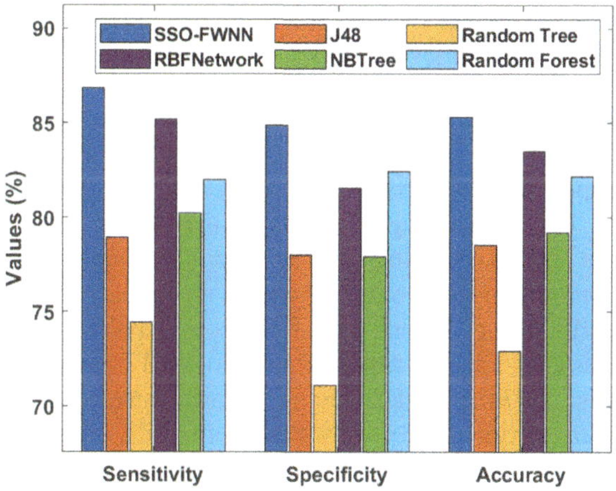

Fig. 5 Heart disease analysis of the proposed SSO-FWNN method

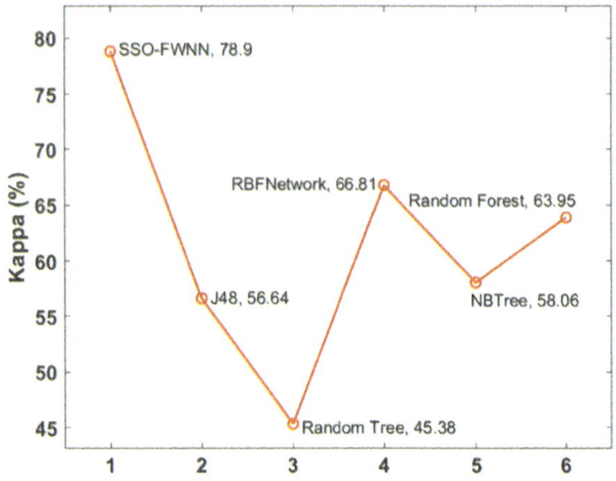

Fig. 6 Kappa analysis of the proposed SSO-FWNN method in terms of sensitivity

NBTree approach attempted to demonstrate moderate results with a sensitivity of 80.24% and specificity of 77.94%. Likewise, the RF model reached considerable outcomes with 81.98% of sensitivity and 82.44% of specificity respectively.

Likewise, the RBFNetwork mechanism accomplished identical and better classifier outcomes with 85.19% of sensitivity and 81.56% of specificity correspondingly. However, the newly developed SSO-FWNN model performed excellent than the compared models with 84.85% of specificity and 86.87% sensitivity correspondingly.

When evaluating the performance of SSO-FWNN model in terms of accuracy and kappa, it is noted that the values should be higher to infer it as excellent performance. Among the models chosen, RT method produced a less classifier outcome with an accuracy of 72.94% and kappa of 45.38% respectively. Similarly, the J48 model attempted to show a better outcome than RT model in terms of accuracy (78.55%) and kappa (56.64%) correspondingly. Followed by, the NBTree approach produced a gradually higher result with an accuracy of 79.21% and kappa of 58.06% respectively. Likewise, the RF technique attained a manageable outcome with an accuracy of 82.18% and kappa of 63.95% correspondingly.

Along with that, the RBFNetwork scheme accomplished a closer and excellent classifier outcome with an accuracy of 83.49% and kappa of 66.81% correspondingly. However, the projected SSO-FWNN model demonstrated effective results in comparison with existing methods by accomplishing an accuracy of 85.32% and kappa of 78.90% correspondingly.

Table 4 and Figs. 7, 8 show the results attained by SSO-FWNN model on the applied Hepatitis dataset. When determining the performance of SSO-FWNN approach in terms of precision and recall, the values must be higher for outstanding performance. Among the models considered, the OlexGA model provided the least classifier outcome with a precision of 4.43% and recall of 20.60% respectively. Likewise, the MLP model attempted to exhibit a reasonable outcome than OlexGA model with a precision of 81.50% and recall of 81.90% correspondingly. On the other end, the DT model attempted to deploy gradual results with a precision of 82.50% and

Table 4 Results of SSO-FWNN method against existing methods for hepatitis dataset

Methods	Precision	Recall	Accuracy	Kappa
SSO-FWNN	84.76	85.43	85.26	61.98
DT	82.50	83.90	83.87	43.60
LR	82.80	83.90	83.87	46.44
MLP	81.50	81.90	81.93	43.56
RBFNetwork	82.60	80.60	80.64	45.93
OlexGA	4.43	20.60	20.64	0

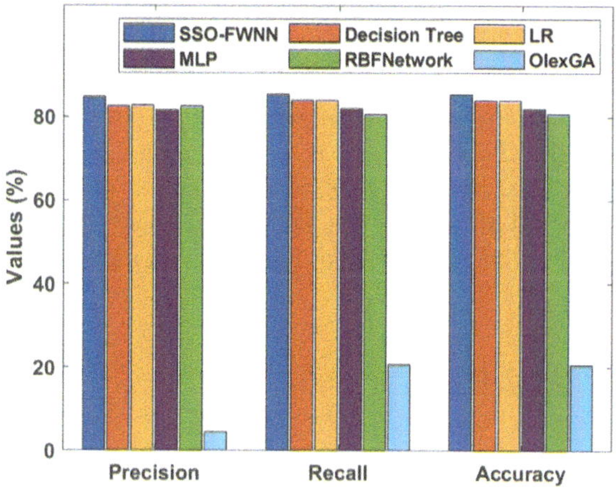

Fig. 7 Hepatitis analysis of the proposed SSO-FWNN method

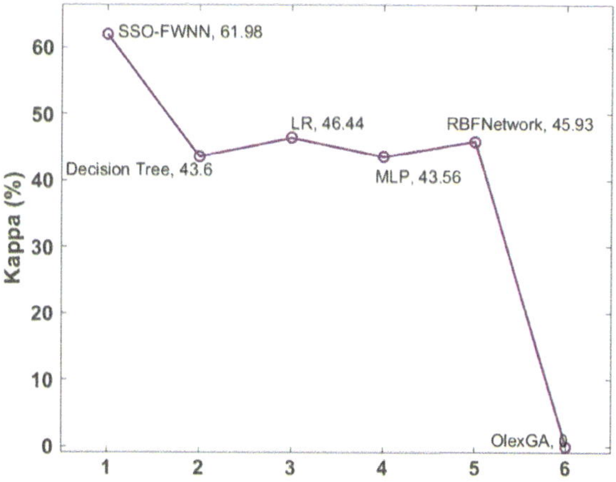

Fig. 8 Kappa analysis of the proposed SSO-FWNN method

recall of 83.90% correspondingly. Along with that, the RBFNetwork model achieved considerable outcomes with a precision of 82.60% and recall of 80.60% respectively. Similarly, the LR model attained the best possible classifier outcome with a precision of 82.80% and recall of 83.90% correspondingly. Hence, the presented SSO-FWNN model illustrated effectual results over existing models with precision being 84.76% and recall being 85.43% correspondingly.

When validating the function of SSO-FWNN model by means of accuracy and kappa, the values should be maximum for efficient performance. Among the techniques considered, the OlexGA model produced the least classifier outcome with an accuracy of 20.64%. Likewise, the RBFNetwork model attempted to show an appreciable outcome than OlexGA model whereas its accuracy was 80.64% and kappa was 45.93%. The MLP model attempted to display reasonable results with an accuracy of 81.93% and kappa of 43.56% correspondingly. In line with this, the DT model achieved a gradual outcome while its accuracy was 83.87% and kappa was 43.60%. Similarly, the LR model accomplished near feasible classifier outcomes with an accuracy of 83.87% and kappa of 46.44% correspondingly. However, the proposed SSO-FWNN model outperformed other methods and exhibited the finest results with an accuracy of 85.26% and kappa of 61.98% correspondingly.

4 Conclusion

The current research work developed an effective SSO-FWNN model for disease diagnosis from medical data. In SSO-FWNN model, SSO method is employed for optimizing the parameters of FWNN to attain maximum classification results. In order to validate the performances of SSO-FWNN model, a series of experiments were conducted on three medical benchmark datasets namely breast cancer, heart disease, and hepatitis. The outcome of the simulations indicates that the SSO-FWNN model achieved maximum outcome on all the applied datasets over existing methods in a significant manner. In future, the performance of the proposed model can further be improved by using clustering and feature selection methodologies.

References

1. Sajda, P.: Machine learning for detection and diagnosis of disease. Annu. Rev. Biomed. Eng. **8**, 537–565 (2006)
2. Yang, X.: Bat algorithm: a novel approach for global engineering optimization. Eng Comput: Int J ComputerAid Eng Softw **29**(5), 464–483 (2010)
3. Rashedi, E., Nezamabadi-pour, H., Saryazdi, S.: GSA: a gravitational search algorithm. Inf. Sci. **179**(13), 2232–2248 (2009)
4. Ghasemi, A., Shayeghi, H., Alkhatib, H.: Robust design of multimachine power system stabilizers using fuzzy gravitational search algorithm. Int. J. Electr. Power Energy Syst. **51**, 190–200 (2013)

5. Sombra, A., Valdez, F., Melin, P.: A new gravitational search algorithm using fuzzy logic to parameter adaptation. In: IEEE Congress on Evolutionary Computation. Cancun, México, pp. 1068–1074 (2013)
6. Beymer, D., Kumair, R., Tanveer, S., Wang, F.: Cardiac disease detection from echocardiogram using edge filtered scale-invariant motion features. In: IEEE Computer Society Conference on Computer Vision and Pattern Recognition Workshops (CVPRW), pp. 162–169 (2010)
7. Kang, S., Hong, S.: A speckle reduction filter using wavelet-based methods for medical imaging application. In: International Conference on Digital Signal Processing (DSP), pp. 1169–1172 (2002)
8. Shahram, E., Shih-Fu, C., Wu, H.: Automatic view recognition in echocardiogram videos using parts-based representation. In: Proceedings of the IEEE Computer Society Conference on Computer Vision and Pattern Recognition (vol. 2, pp. II-2–II-9) (2004)
9. Bailes, D.: The use of the gray level sat to find the salient cavities in echocardiograms. J. Vis. Commun. Image Rep. **7**(2), 169–195 (1996)
10. Dwivedi, Y.K., Hughes, L., Ismagilova, E., Aarts, G., Coombs, C., Crick, T., Duan, Y., Dwivedi, R., Edwards, J., Eirug, A., Galanos, V.: Artificial intelligence (AI): multidisciplinary perspectives on emerging challenges, opportunities, and agenda for research, practice and policy. Int. J. Inform. Manage., 101994 (2019)
11. Banning, A., Behrenbruch, C., Kelion, A., Jacob, G., Noble, J.: A shape-spacebased approach to tracking myocardial borders and quantifying regional leftventricular function applied in echocardiography. IEEE Transact. Med. Imag., 226–238 (2002)
12. Sugioka, K.: Automated quantification of left ventricular function by the automated contour tracking method. J. Cardiovas. Ultrasound Allied Technol., 313–318 (2003)
13. Comaniciu, D., Krishnan, S., Zhou, X.: Robust real-time myocardial border tracking for echocardiography. IEEE Trans. Med. Imaging **23**, 849–860 (2004)
14. Wu, H., Bowers, D., Huynh, T., Souvenir, R.: Echocardiogram view classification using low-level features. In: IEEE 10th International Symposium on Biomedical Imaging: From Nano to Macro. San Francisco, CA, USA, pp. 752–755 (2013)
15. Dowlatshahi, M., Nezamabadi-pour, H.: GGSA: a grouping gravitational search algorithm for data clustering. Eng. Appl. Artif. Intell. **36**, 114–121 (2014)
16. Alneamy, J.S.M., Alnaish, Z.A.H., Hashim, S.M., Alnaish, R.A.H.: Utilizing hybrid functional fuzzy wavelet neural networks with a teaching learning-based optimization algorithm for medical disease diagnosis. Comput. Biol. Med. **112**, 103348 (2019)
17. Cuevas, E., Cienfuegos, M., ZaldíVar, D., Pérez-Cisneros, M.: A swarm optimization algorithm inspired in the behavior of the social-spider. Expert Syst. Appl. **40**(16), 6374–6384 (2013)

Fuzzy with Gravitational Search Algorithm Tuned Radial Basis Function Network for Medical Disease Diagnosis and Classification Model

Denis A. Pustokhin, Irina V. Pustokhina, M. Ilayaraja, and K. Shankar

Abstract The prevalence of diabetes across the globe in the recent decades increased the mortality rate significantly due to improper lifestyle, work style, and absence of physical activities. Being an incurable and chronic disease, if diagnosed from blood glucose level during early stages, Diabetes can be prevented and its harmful effects can be avoided. Several works have been presented based on machine learning and deep learning models in the diagnosis of diabetes. The current research work devises a new diabetes disease diagnosis model using optimization of Radial Basis Function (RBF) with fuzzy-based Gravitational Search Algorithm (FGSA). The proposed FGSA-RBF model intends to determine the optimal structure of RBF Network. Using fuzzy logic in GSA helps in improving the exploration and exploitation capabilities of the proposed model. A detailed experimental analysis was conducted using PIMA Indians Diabetes dataset. The FGSA-RBF model produced superior results in the experiments with maximum precision of 93.56%, recall of 81.02%, accuracy of 83.32%, F-score of 85.23% and kappa value of 65.42%.

Keywords Diabetes · Data classification · Fuzzy logic · Soft computing · Gravitational search algorithm

---

D. A. Pustokhin
State University of Management, Moscow, Russian Federation

I. V. Pustokhina
Plekhanov Russian University of Economics, Moscow, Russian Federation
e-mail: pustohina.iv@rea.ru

M. Ilayaraja
Department of Computer Science and Information Technology, Kalasalingam Academy of Research and Education, Krishnankoil, India
e-mail: ilayaraja.m@klu.ac.in

K. Shankar (✉)
Federal University of Piauí, Teresina 64049-550, Brazil
e-mail: drkshankar@ieee.org

D. Gupta et al. (eds.), *Soft Computing for Data Analytics, Classification Model, and Control*, Studies in Fuzziness and Soft Computing 413,
https://doi.org/10.1007/978-3-030-92026-5_3

1 Introduction

Blood is the essential component in cardiovascular system which is critical for the survival of human beings. Blood can be maintained in a healthy state by consuming vitamin-rich foods. When blood glucose level increases for a prolonged period of time, it results in Diabetes. Increased blood sugar is highly dangerous and results in different health issues. Diabetes occurs among a wide range of age groups that include children, pregnant women, elderly people and even young adults. There are three kinds of diabetes reported so far namely, type 1, type 2, and gestational diabetes. Diabetes provokes the incidence of different health issues in human body such as nerve failure, vision problems, Heart Disease (HD), kidney failure, dental disease, etc. The general signs of diabetes are excess appetite, often feeling thirsty, frequent urination, obesity, and sometimes loss of weight. Typically, Type-2 diabetes (T2D) is said to be a persistent disease. It occurs when there is no sufficient insulin production from pancreas, while the range required differs based on the individual. The primary disease analysis model helps in saving human life through appropriate analysis and medications. Various factors are used by physicians to diagnose the disease. However, there are high chances exist for biased, inappropriate or wrong diagnosis of the disease due to inefficient process or the healthcare professionals. Hence, initial disease prediction model tends to recover the human lives through the prevention of disease development.

Diabetes has been recorded mostly among women compared to men and its prevalence is increasing drastically in the recent years. Iyer et al. [1] provided a classifier with the help of mining approach for diabetes realization. This diabetic detection depends upon Decision Trees (DT) and Naive Bayes (NB) methodologies for pregnant women. The results were estimated with the help of tenfold Cross-Validations (CV) on DT and NB models. Diabetes can be efficiently managed with proper diagnosis when performed without any complications of prediction models. Opper and Winther [2] presented Gaussian Process Classification (GPC) method for disease analysis depending on Machine Learning (ML) approach. Basically, GPC model has three kinds of kernels such as, Linear, Polynomial and Radial. The proposed GPC model was validated for its performance by comparing it with LDA, NB, and QDA methodologies. Accuracy, sensitivity, and specificity of GPC scheme were validated and compared with alternate approaches. The outcomes portrayed the efficiency of ML in diabetes data forecasting [3–8].

Santhanam and Padmavathi [9] projected the K-mean and Genetic Algorithm (GA) for the purpose of diabetes prediction. K-mean technique is mainly applied in noise elimination, while GA model provides active solutions. In this study, the best possible features were categorized under the application of Support Vector Machine (SVM) classifier and offered effectual diagnostic accuracy. Kumari and Singh [10] introduced an approach to perform automated diabetes prediction with the help of Neural Network (NN). Initially, system development and training were processed and the input was provided to the upcoming layer with related bias. Also, Back-Propagation (BP) error was measured and the learning task was performed. Thus, maximum accuracy was achieved in the results.

Samant and Agarwal [11] projected diabetes forecasting method by applying ML models for iris images. This model is mainly applied in the examination of type-2 diabetes with which consistent and valid results can be attained. It depends upon the Region of Interest (ROI) from iris images. Under the application of ROI, statistical, textural as well as Discrete Wavelength Transformation (DWT) can be extracted. Classification, depending on Random Forest (RF) classifier, offers resourceful specificity and sensitivity. Swapna et al. [12] deployed a diabetes prediction method with the help of CNN, CNN-LSTM as well as heart rate signals acquired from ECG. This detection system took the variations of heart rate in ECG signals into account to predict diabetes. CNN-LSTM is mainly utilized to perform automatic detection of irregular heartbeats. The database employed in this approach was divided into training and testing components and high accuracy was obtained with the help of fivefold CV.

Mercaldo et al. [13] developed a ML system to compute disease diagnosis and classification of diabetes. The main goal of this framework was to determine the healthy and diabetes-affected people by applying ML approach. The test, carried out on PIMA dataset of female patients, was derived from UCI ML repository. It is comprised of values present in medical reports. Ijaz et al. [14] employed a hybrid detection approach for type-2 diabetes. Density-relied clustering and over-sampling models were used in balancing the class distribution. Disease classification was performed using RF classifier. The newly-developed method accomplished efficient results by detecting diabetes and hypertension with best accuracy.

Zhu et al. [15] established multiple classification models to detect diabetes at early stages. Marateb et al. [16] projected a hybrid prediction of diabetes with no application of data, relevant to urinary albumin. Such hybrid models provide valid results in terms of sensitivity, specificity, and accuracy. Ganji and Abadeh [17] proposed a ML-based model using Gaussian function to perform the classification of diabetes. A total of three diverse kernels such as Linear, Polynomial and Radial basis was applied in this research work. The model was capable of providing maximum accuracy, when compared with classical models. Beloufa and Chikh [18] presented a hybrid diabetes prediction method with the help of fuzzy classifier and ABC optimization. The performance of the model can be enhanced by the inclusion of mutation operator. The diversity of ACO got improved while no reduction was observed in the superiority of the solution. It got optimized autonomously with Membership Function (MF) and rules from data.

Kumar et al. [19] projected an optimized NN to perform diabetes detection in clinical support system. The disease is diagnosed after the submission of details in the support system. The performance was effectual as it applied ACO model. Chikh et al. [20] provided a mechanism for diabetes detection in which fuzzy and k-nearest neighbor (kNN) were used. It is assumed to be a portion of artificial immune recognition system that classifies the clinical data. The model is capable of providing valid classification accuracy and was verified through tenfold CV. Pérez-Gandía et al. [21] deployed a hybrid detection system for diabetes with the application of parameter hypertension. Noisy data got eliminated by applying DBSCAN technology whereas class distribution was managed under the employment of SMOTE (Synthetic Minority Over-Sampling Technique) and the classification was performed using RF classifier. The experiment was processed on standard datasets and was applied in detecting diabetes and hypertension threats.

The current research article devises a new diabetes disease diagnosis model through the optimization of Radial Basis Function (RBF) with Fuzzy Based Gravitational Search Algorithm (FGSA). The proposed FGSA-RBF model intends to determine the optimal structure of RBFNetwork. Using fuzzy logic in GSA helps in improving the exploration and exploitation capabilities of the model. A detailed experimental analysis was conducted using PIMA Indians Diabetes dataset.

2 The Proposed FGSA-RBF Model

Figure 1 shows the working principle of FGSA-RBF model. As shown in the figure, the diabetes dataset first undergoes preprocessing to remove the unwanted data exists in it. Then, classification process occurs using FGSA-RBF model. This model classifies the data instances into either diabetes or not. The application of FGSA model assists in the optimization of hyper parameters of RBF model.

2.1 Architecture of RBF Model

RBF method is a non-linear classification approach. Using a coefficient, every input value is multiplied and the final outcome is summed. MLP neuron might be a simple linear classifier, but it is composed of non-linear classifiers which are produced by correlating the neurons in the system. From a training set, RBF captures a group of input similarities that shift to identical samples. 'Model' is recorded by all RBFs under the training set. Neurons calculate the Euclidean distance from prototype and input, whenever it is required with an efficient input. If the input is reflected as prototype of class_A over class_B, then it is classified as class_A. The structural method of RBF [22] contains an input vector, RBF neuron as well as output layer with single node for all the class labels as shown in Fig. 2.

Input Vector

It is defined as to-be-classified n-dimensional vector. The whole input vector appears for all the neurons in RBF.

RBF Neurons

'Prototype' is saved by every RBF neuron except the training set. A comparison is made between neuron and the prototype; finally, a value between 1 and 0 is termed

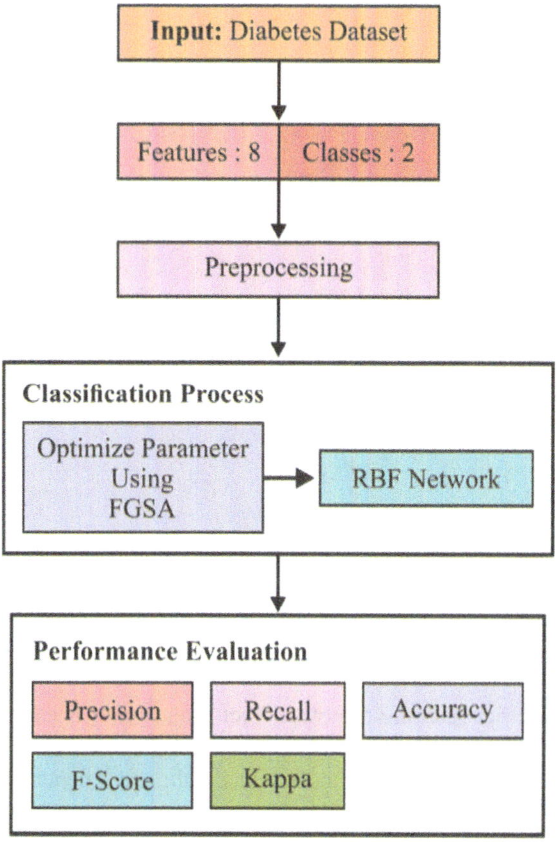

Fig. 1 Block diagram of FGSA-RBF model

as a similarity measure. The result acquired from RBF neuron might be 1, while the same input is at the prototype. However, the simulation outcome gets reduced with an improvement in distance between a prototype and the input. The bell curve implies the structure of a neuron's response. The response value of neuron is named as 'activation' rate.

Output Nodes

A group of nodes is present in the result of the system in which 1 for each label requires a classification task. An input is assigned to a class label containing high score and the classification is proceeded. From every RBF neuron, the values are determined using weighted total of activation rates. A weight value, connected to the output node, is allotted for every neuron in the RBF. The activation of the neurons gets improved with the application of existing weight to overall response. With distinct classes, every output node is calculated and composed with a weight set. A positive weight is offered by an output node to RBF neurons depending on the class label.

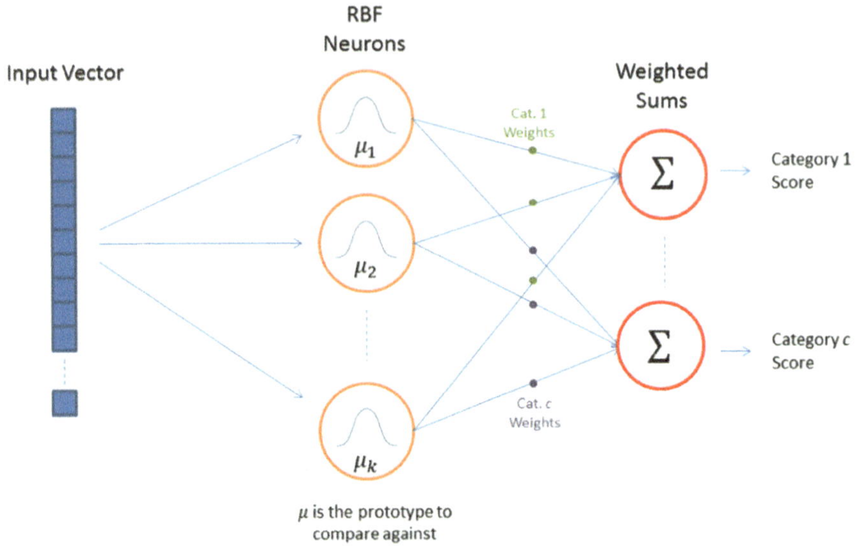

Fig. 2 RBF model

RBF Neuron Activation Function

A similarity value between the prototype vector and input is estimated by all the RBF neurons. The input and output vectors have similar value i.e., '1'. A collection of various feasible similarity functions is selected though it depends upon Gaussian function. Under the application of 1D input, Gaussian expression is depicted in Eq. (1). Assume the average as μ, and input i. It produces a bell curve referred to as centered. The activation function of RBF method is expressed as,

$$\emptyset(i) = e^{-\beta\|i-\mu\|^2} \quad (1)$$

2.2 *Gravitational Search Algorithm*

In order to optimize the parameters of RBF, FGSA is applied. Meta-heuristic approaches resemble biological or physical computation. One of the meta-heuristic models that got evolved from physical laws of GSA was presented by Rashedi et al. [23]. According to populations, it acquires the law of gravity and motion to explore the best solution that has been determined by the masses. These objects are attracted by one another under a gravity force which results in the global movement of objects. Meanwhile the masses are incorporated with the help of direct communication and gravitational force. The agent, with weighted mass, leads to generate better solutions and the movements are slow in comparison with alternate ones. As a result, gravitational and inertial masses are computed with the help of a Fitness Function (FF). In order to define the GSA, a model with N masses is provided in which the position of ith mass is expressed as:

$$\chi_i = \left(X_i^1, \ldots, \chi_i^d, \ldots, X_i^n\right) \quad for \;\; i = 1, 2, \ldots, N \tag{2}$$

where χ_i^d shows the position of ith agent in dth dimension. The mass of an agent is evaluated while the fitness of a recent population is computed as given herewith.

$$M_{oi} = M_{pi} = M_{ii} = M_i, \quad i = 1, 2, \ldots, N, \tag{3}$$

$$\mathrm{m_i(t)} = \frac{\mathrm{fit_i(t) - worst(t)}}{\mathrm{best_t(t) - worst(t)}} \tag{4}$$

$$M_i(t) = \frac{\mathrm{m_i(t)}}{\sum_{j=1}^{N} \mathrm{m_i(t)}} \tag{5}$$

where $M_i(t)$ and $fit_i(t)$ are the mass and fitness values of agent i at time r, correspondingly. In case of minimization problem, $worst(t)$ and $best(t)$ are represented as provided herewith.

$$best(t) = \min_{j \in \{1, \ldots N\}} fit_j(t) \tag{6}$$

$$worst(t) = \max_{j \in \{1, \ldots N\}} fit_j(t) \tag{7}$$

GSA is exploited to resolve the maximization problem which is represented in the Eqs. (8) and (9).

$$best(t) = \max_{j \in \{1, \ldots N\}} fit_j(t) \tag{8}$$

$$worst(t) = \min_{j \in \{1, \ldots N\}} fit_j(t) \tag{9}$$

In order to process the acceleration of an agent, the overall forces from a heavier mass are applied based on the law of gravity. This is then applied in the determination of agent's acceleration by following law of motion. Followed by, velocity and position of an agent are upgraded on the basis of Eqs. (12) and (13)

$$F_i^d(t) = \sum_{j \in K\ best\ j \neq 1} rand_i F_{ij}^d(t) \quad (10)$$

where $randj$ defines the arbitrary value in the range of [0, 1]. Hence, acceleration is expressed as follows.

$$a_i^d(t) = \frac{F_i^d}{M_{ii}(t)} \quad (11)$$

where M_{ii} implies the inertial mass of ith agent. In order to compute the velocity of the agent, a fraction of velocity is included with acceleration.

$$V_i^d(t+1) = rand_i x V_i^d(f) + a_i^d(t) \quad (12)$$

The position of an agent can be evaluated at a given time t, which is added to the velocity at time $t + 1$ as provided herewith.

$$X_i^d(t+1) = x_i^d(r) + \gamma_i^d(t+1) \quad (13)$$

where $rand_i$ and $rand_j$ are two uniformly distributed random values in the range of $[0, 1], \varepsilon$ defines a small value, $R_{ij}(t)$ implies Euclidean distance between the agents i and j and $\|X_i(t) - Xg(t)\|_2$, k best is a collection of first K. agents with leading fitness value and enormous masses allocated to K_0 with decreasing time. In this approach, K_0 implies a set of s which gets reduced to 1 linearly. Then, a gravitational constant G is loaded in starting and later stages which declines after a time and the search accuracy is managed. It is expressed as follows.

$$G(t) = G(G_0, t) \quad (14)$$

A surpassing representation of GSA mechanism is assumed to be the basic principle of this model.

Initially, primary population is provoked which then calculates the fitness of an agent. Thus, the gravitational constant G as well as best and worst of a population are upgraded. Secondly, mass and acceleration of an agent are computed, when the termination criterion is met. Then, more number of iterations are carried out to achieve the perfect solutions; otherwise, it implements the similar procedures of fitness estimation.

Finally, adjustment of this model is utilized i.e., the change of alpha parameter helps in upgrading both G and GSA works effectively. To be specific, G is updated which defines the presentation of a fuzzy model applied to get the measures of alpha and maximize the value of G.

2.3 Fuzzy Logic in Gravitational Search Algorithm to Change Alpha Value

The main theme of improving GSA with FL is acquired from exploration and exploitation strategies in meta-heuristics. In traditional models, a minimum alpha value should be provided in search technique. This is because the technique requires maximum G value to apply force and acceleration majorly, and assist in the exploration of a method [24]. Besides, the final iteration should be provided with maximum alpha value. Because a small value in G helps the method to perform exploitation operation. Hence, a fuzzy system is introduced as given herewith. Here, *iteration* is applied as an input variable, which shows the overall number of iterations that ranged from 0 to 100%. Further, *alpha value* is employed as an output variable. Fuzzy rules are provided as follows.

- "And method" has been utilized as minimum,
- "Or method" has been employed as maximum.

The implication was minimum, and the aggregation was maximum, and a class of defuzzification is said to be a centroid. Figure 3 shows the overall process of FGSA.

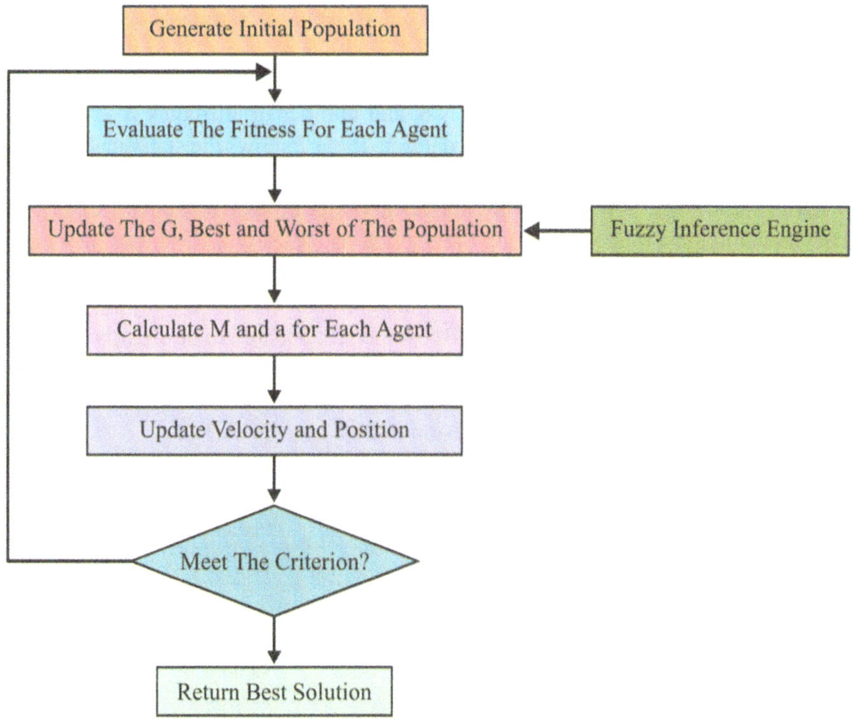

Fig. 3 Flowchart of FGSA

The result is meant to be an alpha value which is granulated as Low, Medium and High. Even though the MF ranges are the lowest and the highest, only positive values must be allocated for alpha. It is computed by an extensive application of alpha value. This MF is a triangular type as defined.

Following is the list of fuzzy rules.

1. *If (Iteration is Low) then (Alpha is Low).*
2. *If (Iteration is Medium) then (Alpha is Medium).*
3. *If (Iteration is High) then (Alpha is High).*

These fuzzy rules are modelled on the basis of lower iterations, where an alpha value has to be low since it requires maximum gravity and acceleration to process the search space exploration. Additionally, the final iterations require maximum alpha values to decrease the agents of gravitation and acceleration and to make use of a search space. A similar procedure is followed in case of medium iterations, since a median amount of gravitation is presented for the agents.

3 Performance Validation

In order to validate the effectiveness of FGSA-RBF model, extensive experimental analyses were carried out using PIMA Indian Diabetes dataset [25]. The information related to the dataset is provided in Table 1. As per the table values, the number of instances is 768, number of attributes is 8, and number of attributes is 2. Figure 4 shows the frequency distribution of the dataset. Table 2 shows the statistical analysis of Pima Indian Diabetes Dataset.

Table 3 and Figs. 5, 6 and 7 show the results of comparative analysis of FGSA-RBF model against other models on the applied dataset. Figure 5 shows the results for precision analysis of FGSA-RBF model against the compared methods. The figure states that the DT model offered an insignificant classification outcome with a minimum precision of 81.40%. LogitBoost model appeared to be slightly effective over DT since it offered a precision of 84.60%. LR model resulted in a moderate and acceptable precision of 88%. Afterwards, the Voted Perceptron model accomplished a competitive performance with a precision of 88%.

Table 1 Dataset description

Description	Pima Indian diabetes
No. of instances	768
No. of attributes	8
No. of class	2
Positive samples (%)	34.90%
Negative samples (%)	65.10%
Data sources	[22]

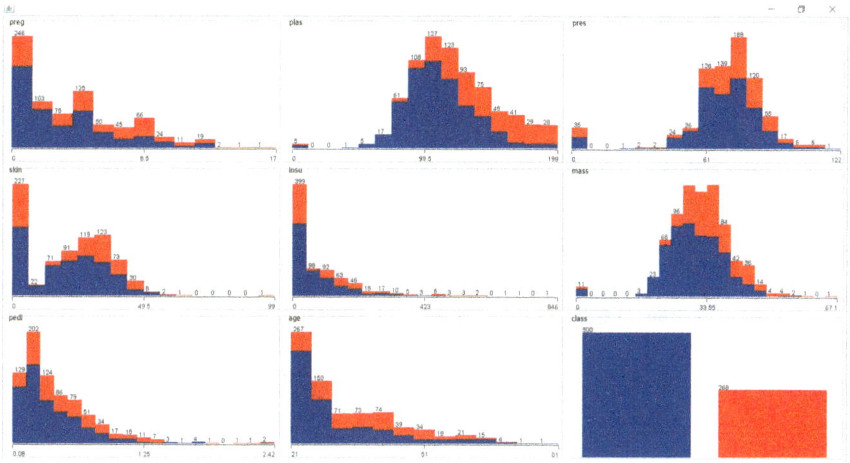

Fig. 4 Frequency distribution of pima Indian diabetes dataset for all attributes

Table 2 Statistical analysis of pima Indian diabetes dataset

Attributes	Description	Type	Min	Max	Mean	Std. Dev
Prig	Number of times pregnant	Numerical	0	17	3.84	3.37
Plas	Plasma glucose concentration	Numerical	0	199	120.89	31.97
Pres	Diastolic blood pressure	Numerical	0	122	69.10	19.36
Skin	Triceps skin fold thickness	Numerical	0	99	20.54	15.95
Insu	2-h serum insulin	Numerical	0	846	79.79	115.24
BMI	Body mass index	Numerical	0	67.1	31.99	7.88
Pedi	Diabetes pedigree function	Numerical	0.078	2.42	0.472	0.331
Age	Age	Numerical	21	81	33.241	11.76

Table 3 Performance evaluation of different classifiers with proposed FGSA-RBF method

Classifiers	Precision	Recall	Accuracy	F-score	Kappa
FGSA-RBF	93.56	81.02	83.32	85.23	65.42
LR	88.00	79.27	77.21	83.41	47.34
Voted perceptron	92.40	68.04	66.79	78.37	13.52
LogitBoost	84.60	77.61	74.08	80.95	40.66
DT	81.40	79.02	73.83	80.19	41.64

In this scenario, FGSA-RBF model outperformed all the earlier methods and achieved a maximum precision of 93.56%. The figure implies that the Voted Perceptron method provided ineffective classification outcome with low recall of 68.04%. Followed by, the LogitBoost approach exhibited a better result than Voted Perceptron

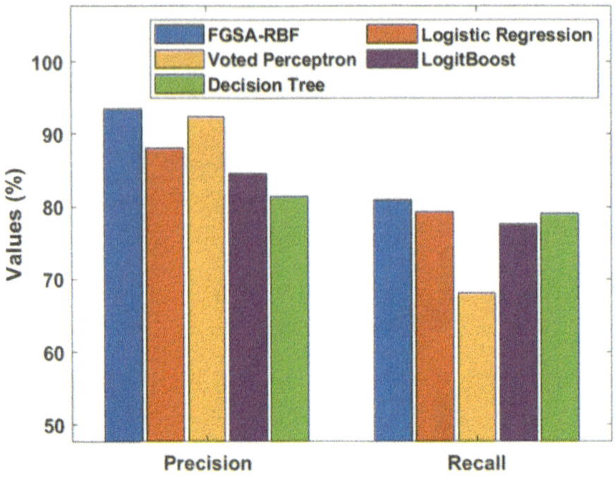

Fig. 5 Comparative analysis of existing model in terms of precision and recall

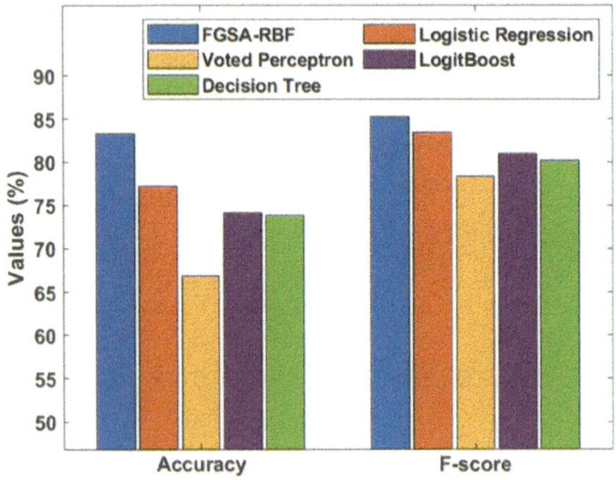

Fig. 6 Comparative analysis of existing models in terms of accuracy and F-score

and provided a recall of 77.61%. Then, the DT model attained a gradual and acceptable recall of 79.02%. Later, the LR approach resulted in a competing performance with a recall of 79.27%. Finally, the FGSA-RBF model performed quite well than the previous techniques and achieved a high recall of 81.02%.

Figure 6 shows the results for accuracy analysis of FGSA-RBF model with related methods. The figure portrays that the Voted Perceptron model achieved an irregular classification outcome with its accuracy being 66.79%. Next, the DT model attained a considerable result than Voted Perceptron by accomplishing an accuracy of 73.83%. Followed by, the LogitBoost model concluded with a moderate and applicable accuracy of 74.08%. LR model accomplished an equal performance and an accuracy of 77.21%. However, the proposed FGSA-RBF model exhibited an

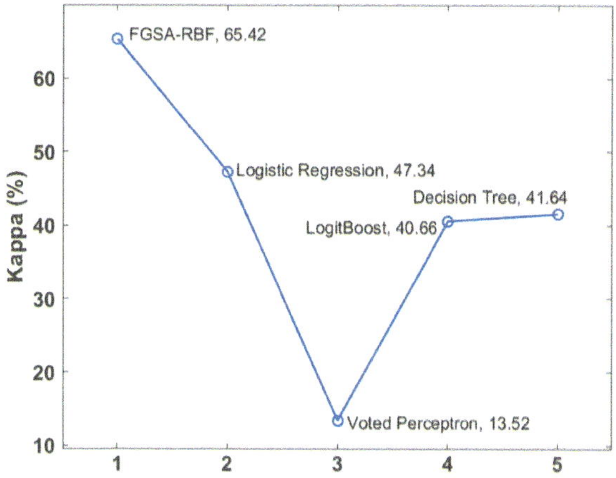

Fig. 7 Comparative analysis of existing models in terms of kappa

outstanding performance than existing methods and achieved a maximum accuracy of 83.32%.

The figure reveals that the Voted Perceptron model offered immaterial classification outcome with the lower F-score of 78.37%. Next, the DT model showcased a better outcome than Voted Perceptron and generated an F-score of 80.19%. Then, the LogitBoost model provided a considerable and satisfactory F-score of 80.95%. Later, the LR model resulted in a competing performance with an F-score of 83.41%. In this scenario, FGSA-RBF model accomplished extra-ordinary performance against previous methods and achieved the highest F-score of 85.23%.

Figure 7 shows the results of kappa analysis achieved by FGSA-RBF model, in comparison with alternate methods. The figure portrays that the Voted Perceptron model provided an ineffective classification outcome with a low kappa of 13.52%. Next, the LogitBoost model showcased slightly moderate performance above Voted Perceptron and produced a kappa value of 40.66%. Followed by, the DT model attained a gradual and convenient kappa of 41.64%. LR model produced a competing result with a kappa of 47.34%. However, FGSA-RBF model obtained optimal results than the existing methods and accomplished a maximum kappa of 65.42%. The above-mentioned tables and figures infer that the FGSA-RBF model is a superior performer and produced excellent results in terms of precision (93.56%), recall (81.02%), accuracy (83.32%), F-score (85.23%) and kappa value (65.42%).

4 Conclusion

The current research article developed an efficient RBF-FGSA technique to diagnose diabetes. The proposed RBF-FGSA model involves preprocessing, parameter optimization and classification processes. FGSA-RBF model focused on determining the optimal structure of RBFNetwork. Fuzzy logic was utilized in GSA which in turn helped in improving the exploration and exploitation capabilities. The researcher conducted detailed experimental investigations using PIMA Indians Diabetes dataset. In the experiments, the proposed FGSA-RBF model accomplished superior results with a maximum precision of 93.56%, recall of 81.02%, accuracy of 83.32%, F-score of 85.23% and kappa value of 65.42%. Hence, the model has been established for its supremacy.

Acknowledgements Dr. K. Shankar sincerely acknowledge the financial support of RUSA–Phase 2.0 grant sanctioned vide Letter No. F. 24-51/2014-U, Policy (TNMulti-Gen), Dept. of Edn. Govt. of India, Dt. 09.10.2018.

Conflict of Interest The authors declare that they have no conflict of interest. The manuscript was written through contributions of all authors. All authors have given approval to the final version of the manuscript.

References

1. Iyer, A., Jeyalatha, S.. Sumbaly, R.: Diagnosis of diabetes using classification mining techniques (2015). arXiv:1502.03774
2. Opper, M., Winther, O.: Gaussian processes for classification: mean-field algorithms. Neural Comput. **12**, 2655–2684 (2000)
3. Gadekallu, T.R., Khare, N., Bhattacharya, S., Singh, S., Maddikunta, P.K.R., Ra, I.H., Alazab, M.: Early detection of diabetic retinopathy using PCA-firefly based deep learning model. Electronics **9**(2), 274 (2020)
4. Dutta, S., Manideep, B.C., Basha, S.M., Caytiles, R.D., Iyengar, N.C.S.N.: Classification of diabetic retinopathy images by using deep learning models. Int. J. Grid Distrib. Comput. **11**(1), 89–106 (2018)
5. Das, H., Naik, B., Behera, H.S.: Classification of diabetes mellitus disease (DMD): a data mining (DM) approach. In: Progress in Computing, Analytics and Networking (pp. 539–549). Springer, Singapore (2018)
6. Ghosh, R., Ghosh, K., Maitra, S.: Automatic detection and classification of diabetic retinopathy stages using CNN. In: 2017 4th International Conference on Signal Processing and Integrated Networks (SPIN) (pp. 550–554). IEEE (2017, February)
7. Adal, K.M., Van Etten, P.G., Martinez, J.P., Rouwen, K.W., Vermeer, K.A., van Vliet, L.J.: An automated system for the detection and classification of retinal changes due to red lesions in longitudinal fundus images. IEEE Trans. Biomed. Eng. **65**(6), 1382–1390 (2017)
8. Chaki, J., Ganesh, S.T., Cidham, S.K., Theertan, S.A.: Machine learning and artificial intelligence based Diabetes Mellitus detection and self-management: a systematic review. J. King Saud Univ. Comp. Inf. Sci. (2020)
9. Santhanam, T., Padmavathi, M.: Application of k-means and genetic algorithms for dimension reduction by integrating SVM for diabetes diagnosis. Procedia Comput. Sci. **47**, 76–83 (2015)
10. Kumari, S., Singh, A.: A data mining approach for the diagnosis of diabetes mellitus. In: Intelligent Systems and Control (ISCO), 2013 7th International Conference on, IEEE, pp. 373–375 (2013)
11. Samant, P., Agarwal, R.: Machine learning techniques for medical diagnosis of diabetes using iris images. Comput. Methods Programs Biomed. **157**, 121–128 (2018)
12. Swapna, G., Kp, S., Vinayakumar, R.: Automated detection of diabetes using CNN and CNN-LSTM network and heart rate signals. Procedia Comput. Sci. **132**, 1253–1262 (2018)

13. Mercaldo, F., Nardone, V., Santone, A.: Diabetes mellitus affected patients classification and diagnosis through machine learning techniques. Procedia Comput. Sci. **112**, 2519–2528 (2017)
14. Ijaz, M., Alfian, G., Syafrudin, M., Rhee, J.: Hybrid prediction model for type 2 diabetes and hypertension using DBSCAN-based outlier detection, synthetic minority over sampling technique (SMOTE), and random forest. Appl. Sci. **8**, 1325 (2018)
15. Zhu, J., Xie, Q., Zheng, K.: An improved early detection method of type-2 diabetes mellitus using multiple classifier system. Inf. Sci. **292**, 1–14 (2015)
16. Marateb, H.R., Mansourian, M., Faghihimani, E., Amini, M., Farina, D.: A hybrid intelligent system for diagnosing microalbuminuria in type 2 diabetes patients without having to measure urinary albumin. Comput. Biol. Med. **45**, 34–42 (2014)
17. Ganji, M.F., Abadeh, M.S.: A fuzzy classification system based on ant colony optimization for diabetes disease diagnosis. Expert Syst. Appl. **38**, 14650–14659 (2011)
18. Beloufa, F., Chikh, M.A.: Design of fuzzy classifier for diabetes disease using modified artificial bee colony algorithm. Comput. Methods Programs Biomed. **112**, 92–103 (2013)
19. Kumar, M., Sharma, A., Agarwal, S.: Clinical decision support system for diabetes disease diagnosis using optimized neural network. In: Engineering and Systems (SCES), 2014 Students Conference on, IEEE, pp. 1–6 (2014)
20. Chikh, M.A., Saidi, M., Settouti, N.: Diagnosis of diabetes diseases using an artificial immune recognition system2 (AIRS2) with fuzzy k-nearest neighbor. J. Med. Syst. **36,** 2721–2729 (2012)
21. Pérez-Gandía, C., Facchinetti, A., Sparacino, G., Cobelli, C., Gómez, E., Rigla, M., de Leiva, A., Hernando, M.: Artificial neural network algorithm for online glucose prediction from continuous glucose monitoring. Diabetes Technol. Ther. **12,** 81–88 (2010)
22. https://mccormickml.com/2013/08/15/radial-basis-function-network-rbfn-tutorial/
23. Rashedi, E., Nezamabadi-Pour, H., Saryazdi, S.: GSA: a gravitational search algorithm. Inf. Sci. **179**(13), 2232–2248 (2009)
24. González, B., Valdez, F., Melin, P., Prado-Arechiga, G.: Fuzzy logic in the gravitational search algorithm for the optimization of modular neural networks in pattern recognition. Expert Syst. Appl. 42(14), 5839–5847 (2015)
25. https://www.kaggle.com/uciml/pima-indians-diabetes-database

Optimal Neutrosophic Rules Based Feature Extraction for Data Classification Using Deep Learning Model

V. Gomathy, T. Jayasankar, M. Rajaram, E. Anna Devi, and S. Priyadharshini

Abstract Classification issues and their relevant solving techniques constitute one of the fields in machine learning. It is a widely used technique in different domains including data mining, communications technology, medicine, science, and so on. The problems experienced in the current system are maximum computational time and feature selection model. To overcome these challenges, an innovative classification process is proposed in this paper. The researchers initially used Kernel Principle Component Analysis (KPCA) model to reduce the features of the selected database. The reason behind the selection of this dimensional approach is to manage the computational time. For better classification process, an optimal rule was generated in Neutrosophic Set (NS) and for this rule optimization, Oppositional-based Elephant Herd Algorithm (OEHA) was used. After selecting the features, Deep Learning (DL) approach was introduced to classify the data. The main advantage of DL is its time-consuming phases of learning practice. The implementation results in terms of precision, recall, and accuracy was found to be 95.56%, 85.23%, and 90% respectively when compared to other classification and feature selection techniques.

V. Gomathy (✉)
Department of Electrical and Electronics Engineering, Sri Krishna College of Engineering and Technology, Coimbatore, India
e-mail: gomathyv@skcet.ac.in

T. Jayasankar
Department of Electronics and Communication Engineering, University College of Engineering, BIT Campus, Anna University, Tiruchirapalli, India

M. Rajaram
Department of Electronics and Communication Engineering, Park College of Engineering and Technology, Coimbatore, India

E. A. Devi
Department of ECE, Sathyabama Institute of Science and Technology, Chennai, India
e-mail: annadevimit@gmail.com

S. Priyadharshini
Department of Electronics and Communication Engineering, Mohamed Sathak A.J College of Engineering, Chennai, India
e-mail: ece.priyadharshini@msajce-edu.in

D. Gupta et al. (eds.), *Soft Computing for Data Analytics, Classification Model, and Control*, Studies in Fuzziness and Soft Computing 413,
https://doi.org/10.1007/978-3-030-92026-5_4

Keywords Data classification · Feature selection · Optimization · Neutrosophic set (NS) · Deep learning (DL)

1 Introduction

Classification techniques are assigned based on data features to predefine the data under groups. The most commonly used algorithm in data mining is 'classification' and 'clustering'. In order to make predictions for the dataset considered, these algorithms work on the basis of dataset dimensions [1], notwithstanding, when the dimensionality of datasets gets expanded, it creates an impact on the execution of clustering and classification strategies too [2]. Based on the training set, classification model is built and employed to data point in the validation set [3]. Further, high dimensional datasets create a wide range of inconveniences, for example, consuming substantial time for model development, excess data, and debased performances which troubles the data analyses process [4]. In other terms, classification or regression models that contain repetitive features tend to result in degraded performance with the enclosure of redundant or even contradictory information [5]. By leveraging feature selection and optimization techniques, such discriminative features can be identified followed by the removal of redundant ones [6]. Feature selection is utilized to avoid dimensionality issues and to reduce the computational weight [7]. When dimensionality is reduced, it disposes of the data without realizing its significance [8].

The determination of 'feature subset' from each conceivable feature subset is taken into account as a combinatorial [9] optimization issue. It is a thorough pursuit in expansive arrangement space that can be handled using Meta heuristic algorithm [10]. Feature selection, in supervised learning, is an effortless process. This is because one can simply choose what needs to be included based on the knowledge about class name and in accordance with the class label [6]. The interval Neutrosophic Set (NS) represents uncertain data and supports binary classification [11]. NS is considered as a section of Neutrosophic hypothesis which gets intrigued by the examination of cause, nature, extent of neutralities, and their associations with different ideational spectra [12]. The classification method has been introduced based on rules and features [13]. Classification is a procedure of categorizing data for its best and efficient use based on prerequisites. It has two stages such as 'learning process', the first stage, and 'classification process', the second stage [12]. The execution of the classifiers may get deteriorated due to over fitting issues. In the meantime, the learned models diminish their interpretability as they become more perplexed. Finally, both speed and the effectiveness of the algorithms [14] get improved by the herding behavior of elephants. Being social animals, elephants live in groups and form tribes in general. Elephants inhabit a place in which different families live together under the authority of a matron. Male elephants, on the other hand, stay single and leave their family gathering alike which grows up in the optimization process too [15].

The major contribution of this manuscript is the classification of data based on optimal NS process and feature reduction approach. The novelty of the proposed model is its data classification and reduction in attribute information, followed by NS generation. Further, the other fascinating feature is that the study considers OEHA to optimize the generated rules for classification process. The paper is organized as follows, Sect. 2 discusses the existing literature and Sect. 3 discusses the research gap found in the literature and the current system of data classification. The proposed methodology and the techniques are determined under Sect. 4; the implementation outcomes are discussed under Sect. 5 and the work is concluded along with the suggestion for future work in Sect. 6.

2 Literature Review

Preprocessing of the data occurs in data reduction system prior to the training stage of classifiers. A novel data reduction method was proposed in the literature depending on k-means clustering algorithm. Data reduction systems have been widely utilized earlier. However, with regards to KNN classification, a lethargic classifier, which works by specifically utilizing a training dataset as opposed to model building was developed by Ougiaroglou et al. [16]. In 2018, Dong et al. [17] developed a method called Improved Binary Genetic Algorithm with Feature Granulation (IBGAFG) has been employed to choose the critical features. At that point, an Improved Neighborhood Set with Sample Granulation (INRSG) was introduced under distinct granular ranges that additionally enhance the nature of the element subset. The ideal granularity parameters were adaptively found as per the criticism put forth by the classification accuracy. The chosen features sustained Multiclass Support Vector Machine (M-SVM) in the classification of definite citrus disease. The presented strategy was tested with citrus disease image gallery dataset, combined dataset (Database combining village plants and citrus images infested with scale), and the image gathered by Sharifa et al. [18] from the database. The nature of features affects the execution of various-leveled learning approaches in the classification of patient records. So Franklin Vinod et al. mentioned that the need of the hour is to separate discriminative features so as to train a progressive classifier [19]. The neutrosophic rationale was adjusted to indicate different sorts of information. GA was utilized to refine the created neutrosophic rules. In 2016, Basha et al. [20] proposed a 'hereditary neutrosophic rule-based classification' against the neutrosophic control-based classification method. The two classifiers were estimated using three data sets. The genetic neutrosophic method achieved a normal accuracy of 98.39%, whereas it was 94.78% in case of the compared neutrosophic method. A metaheuristic algorithm, based on fish behavior underwater and EHO, was proposed in the research work conducted earlier. The wrapper approach of EHO with SVM is utilized in finding the element subset with the least number of features. EHO algorithm was used to fine-tune the SVM parameters and to select the most discriminant feature. The knowledge was gained from discovery tests conducted upon various human activities. The results demonstrated the supremacy of the proposed approach over existing methods, on a similar dataset, with respect to classification accuracy defined earlier by Nalluri et al. in 2017 [21] and Kilany et al. in 2017 [22].

3 Research Gap in Current System

As of late, numerous analysts and experts have been utilizing few machine learning techniques in data classification [20]. A vast majority of the papers have actualized a few medical data classification strategies, for example, neural system, SVM, KNN, and so on. These methods demonstrate distinct levels of accuracy on different databases. The issue of learning an 'optimum decision tree' is known to be NP-finished. Thus, DT learning algorithm is based on heuristic algorithm [20]. For example, in greedy algorithm, local optimal decisions are taken at every hub which is very unintuitive. It is difficult to perceive the reason behind the inclination of the request connection towards a sentence known [17] to be valid, though with extraordinary indeterminacy (imprecision, error, etc.). In the current paper, some classification strategies were utilized to categorize the medical data [21]. But some particular issues are identified with optimization algorithms like high computational time, unclear identification of the attributes or features during training, and so on. So, in order to overcome these issues, the proposed strategy is executed.

4 Methodology of the Proposed Work

In data mining research model, data classification is a challenging task to accomplish. In the proposed work as shown in Fig. 1, data classification is performed using an innovative approach called 'Neutrosophic Set (NS) and Machine Learning Classifier (ML)'. Initially, the data is collected for preprocessing when the null values and unwanted attribute information are removed. Then, the preprocessed data is assigned to feature reduction algorithm (KPCA) and NS rule generation process. This KPCA algorithm helps in selecting useful attributes or features. The main purpose of attribute reduction is to reduce computation time and storage space. Once the feature reduction is modeled, NS is generated. Being a rule-based system, neutrosophic logic is employed as a tool to represent distinct forms of knowledge about the classification model. All the rules are characterized by a 'string' and are handled as an individual.

After NS rule generation, an optimization technique is utilized to optimize the NS rules generated for data classification model. In this study, the researchers used OEHA optimization technique. Finally, DL approach is considered to classify the data with the help of optimal NS rules. A detailed discussion of the proposed model is given in the sections below and the proposed model was implemented in MATLAB platform.

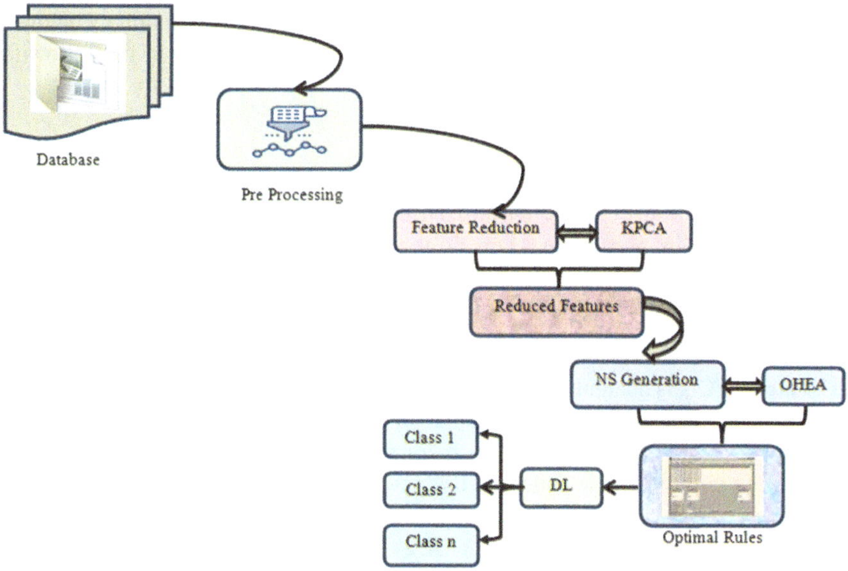

Fig. 1 Block diagram for the proposed model

4.1 Preprocessing: Removal of Missing Information

In the proposed data classification model, various databases were used for the classification of the data. The database contains structured, unstructured, and missing data which are characterized by their incorrectness, inconsistency, redundancy, sparseness, and inexactness.

- Handling missing values and null values in the database; generally, all the attributes of the numerical values vary in the range of (0, 1).
- If the removal of instances, with missing values, gets delayed after attribute selection step, this will permit additional instances to be accessible for testing and training subsets. However, this procedure improves the results as well [23].
- If the removal of occurrences is excused with missing qualities, preprocessing stage gets delayed until the attribute determination process.

4.2 KPCA for Feature Reduction

Dimensionality reduction denotes the selection of an appropriate subset of features or attributes among the existing features and combined features. This is executed by redesigning the feature sets. The proposed KPCA model removes the redundant features present in real database. Feature selection algorithms can be grouped under three categories such as filters, wrapper, and embedded. The current study considered wrapper algorithm which functions mainly based on data classification techniques. The steps involved in KPCA dimension reduction are discussed in the following section.

KPCA procedure

In the proposed model, the features are reduced with the help of KPCA to generate the NS. It is the most commonly utilized non-parametric technique that separates significant data, by presenting a new arrangement of highlights created on another vector space. A reduced arrangement of features, chosen in this space, is anticipated back to the original [24] space. However, the vectors in the first space with base edges, to these reduced vectors, are considered as the final features.

PCA steps: PCA is a latent variable model and characterizes a generative procedure for every datum. Assume the database as $M = \{M_1, M_2, \ldots.M_d\}$, a latent variable related with every datum point by g_d, a connection between the latent variable and information point in whole database. It is scientifically depicted as follows.

$$M_d = R.g_d + \alpha_d \tag{1}$$

where R represents a matrix which identifies the linear relationship among the latent-space, data space, and noise values. $\alpha_d \in noise$ is taken as an independent sample from a spherical Gaussian distribution with mean zero.

$$C(\alpha_d) = D(\alpha_d|0, \beta^{-1}, M) \tag{2}$$

The covariance function of Gaussian distribution is calculated using the equation given below.

$$Con = \frac{1}{D}\sum(g_d - \delta)(g_d - \delta)^T \tag{3}$$

$$\text{Then } \delta = \frac{1}{D}\sum_{n=1}^{D} g_d \tag{4}$$

PCA model is used as a compelled Gaussian thickness function. The benefit of this function is to decrease the measurement of highlights. It has the highest probability measures for parameters related to covariance framework in view of the segments.

Kernel function

The feature reduction procedure of KPCA kernel function was utilized for upgradation. It works by mapping the input data g into a high or unbounded dimension space and after constructing linear algorithms in the component space to execute nonlinear partners in the input space. From covariance lattice model, the kernel is depicted as $K(g, g_1)$ which denotes the internal item between a combination of various information points, whereas the function is as follows.

$$M_d = R * K(g_d, g_d) + \alpha_d \tag{5}$$

The study evaluated Eigenvalues and matrix of the dimension reduction model. KPCA found the direct projection that boosts the difference in the anticipated space. Let g be a q-dimensional latent variable. The observed variable k is then characterized as a direct change of extra noise. Here $R \to$ d × q denotes the linear transformation matrix, whereas $\alpha \to$ d measurement vector enables 't' to have a non-zero mean. The normal perspective of KPCA maps from higher dimension information space to lower dimension latent space. Against this arrangement, KPCA structure depends on mapping from a latent space to information space.

4.3 Neutrosophic Set (NS) Generation

Neutrosophy is a moderately new branch of philosophy with old roots and it takes care of neutralities and their co-operation with various ideational spectra in addition. For powerful data classification model, NS is generated to set the rules; an interim neutrosophic set is an occasion of neutrosophic set, which is summed up from the idea of traditional set. NS procedure, under three occasions such as truth membership, indeterminacy membership, and false membership i.e., $\{T, I\ and\ A\}$ postulates the functions as free and are shown in Fig. 2. In few uncommon cases, they can be relied upon [25]. The lessened component subsets are given to NS membership function and the rules are created with the help of OEHA. Indeterminacy membership relies upon both truth and false memberships.

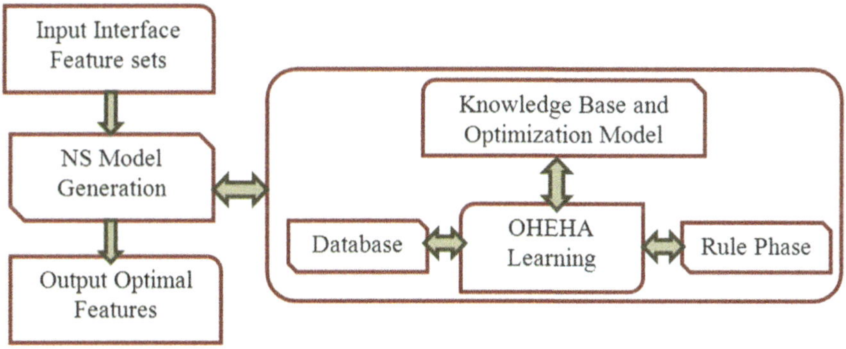

Fig. 2 NS rules and membership function

(i) **Data initialization**

KPCA feature medical data is initialized from the range of [0, 1]. This is because, after initialization only, it is possible to generate NS rules. The function is $NS = \{fe_1, fe_2, ...fe_n\}$.

(ii) **NS probability calculation**

It is a speculation of traditional probability in which the shot which is set at first is valid; the time shifts in the subset of undetermined qualities; likewise, false proclamation is additionally assessed. The speculation of uncertain probability, which is an interim esteemed conveyance function, and the NS probability are characterized by the following equation.

$$p(F) = (m(F), m(NS\_F)) \tag{6}$$

Here NS_F are neutral values, p is the function that satisfies the property of null set and is countable additively. NS_F denotes the measure of determinate portion whereas F denotes the measure of indeterminate portion. From probability calculation, the subset needs to be found for NS rule and membership function process.

(iii) **Rule and Membership function generation**

Memberships functions are intended for each one of the factors i.e. input and output factors, with the end goal that there is no overlapping between two membership functions. The neutrosophic deception part utilizes the preparing set which is planned simultaneously since it improved the situation indeterminacy segment in spite of the tallness of each membership function being {0, 1}. The membership function computation in light of the probability values is characterized as follows

$$M = \{F(T_M(F), I_M(F), A_M(F)\} \tag{7}$$

The range of these three membership functions is as follows.

$$T_M : F \to [0, 1] \wedge \ and \ I_M : F \to [0, 1] \wedge \ and \ A_M : F \to [0, 1] \tag{8}$$

Among the membership functions, the feature set is truth and indeterminacy, whereas the falsehood values referred to neutrosophic set, neutrosophy, neutrosophic statistics, and neutrosophic logic.

The initial population is generated in neutrosophy based on 'If–then' rules. Each rule is framed by a string as in Michigan approach whereas each string is dealt with, like an individual. The populace of people is denoted as a single ruleset. The rule set is not straightforwardly enhanced due to which the execution of rule sets is also not used in the advancement of rules. The example rules of NS process are given herewith [26].

Isolet Database

If speaker 1 is [low, Indeterminacy low, false low], If speaker 1 is [low, false low, 0], speaker 2 [1, 0, Indeterminacy low] and if speaker 2 is [1, 1, indeterminacy low], etc. Then identify the class.

Musk Database

If molecules 1 are [0, indeterminacy medium, indeterminacy high,], If molecules 1 [1, indeterminacy high, 0] and If a molecule 2 is [0, 0, indeterminacy high], If a molecule 2 is [1, 0, indeterminacy low], etc. Then find the class.

Hill-Valley Database

If X1 is [1, Indeterminacy low, false 0], If X1 is [high, false low, 0], X3 [1, 0, Indeterminacy medium] and if alcohol is [0, 1, indeterminacy high], etc. Then identify the class.

In neutrosophic rule-based classification framework, every factor in each rule has three segments such as individual level of truth, level of indeterminacy, and level of deception. At that point, after checking for repetition, the rule is added to preparation rule. For better classification of work, NS rules are to be enhanced with the help of OEHA. This procedure and the ventures of rule optimization are discussed below.

4.4 Oppositional Elephant Herding Algorithm (OEHA)

EHA is framed on the basis of herding behaviors of elephants. All the clans are driven by an authority who symbolizes the optimal arrangement of a clan. The opposition Based learning method ensures the examination of interesting and inverse applicant solutions in search space. At the same time, the model also surveys the best hopeful solutions while the development procedure is ahead of time. The flowchart for ideal rule flowchart is illustrated in Fig. 3 and the steps are discussed below.

Step 1: Initialize the Rules

All the generated rules are initialized in NS process and each rule is considered as one elephant R_E in total clans, C_T. The positions of ith elephant, $i = 1, 2, \ldots\ldots R_E$ in jth clan, $i = 1, 2, \ldots\ldots C_T$ is denoted by α_{ij}. The position of the proposed model is given by the equation given below.

$$\beta_{new,i,j} = \beta_{ij} + \beta\left(\beta_{best,j} - \beta_{ij}\right) * R \tag{9}$$

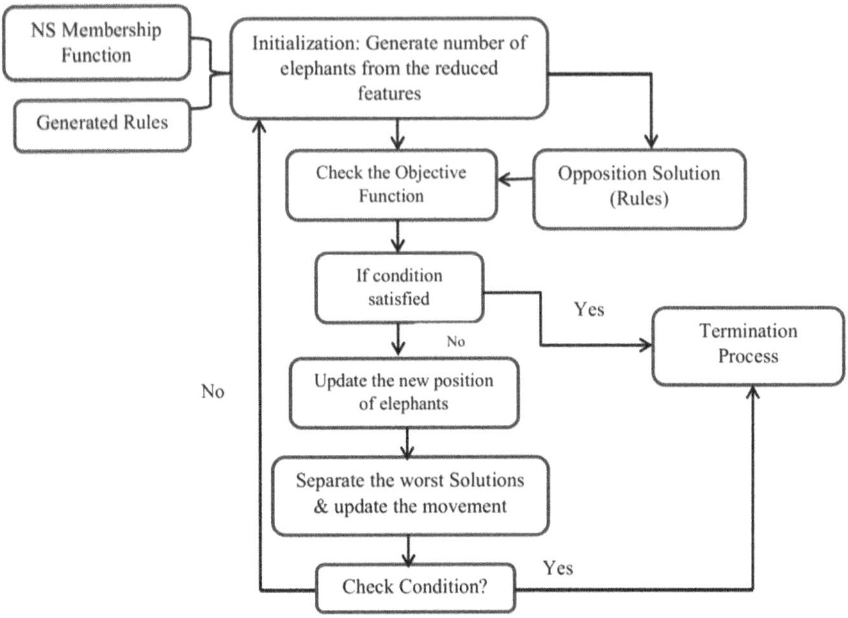

Fig. 3 Flowchart for OEHA

Here $\beta_{new,i,j}$ represents the upgraded locatidesign which comprise inputon, α symbolizes the scale factors that determine the effect of matriarch upon clans like α within [0, 1].

Step 2: Oppositional Process

All the clans are driven by a female authority who symbolizes the best arrangement of the tribe [22]. The opposition Based learning method ensures the examination of special and inverse candidate arrangements in search space. At the same time, the model also surveys the best candidate solutions, while the development procedure is ahead of time. Here, Opposition Function is denoted as follows.

$$O_i^* = U_i + V_i - O_i \tag{10}$$

Based on the above equation, the opposing solution is generated along with fitness values.

Step 3: Objective Function

The proposed optimal rule selection process considered the 'error rate' as a fitness function i.e., activation function which is calculated using confusion matrix tables of the classification model.

$$Objective\ Function = Min\left(\frac{Number\ of\ WP\ and\ WN}{Number\ of\ CP, CN\ and\ WP, WN}\right) \tag{11}$$

Here WP and WN are wrong positive and wrong negative, whereas CP and CN are correct positive and correct negatives. Equation 11 minimizes the classification error rate achieved in the evolutionary training process.

Step 4: Elephant position and Movement Updating

The new updated positions are selected based on the fitness evaluation. The location upgrade for the optimal fit in clan can be shown in Eq. (12).

$$\beta_{best} = \delta \times \beta_{center,j}, \text{ Here } \beta_{center,j} = \frac{1}{T_c}\sum_{i=1}^{n}\beta_{ij} \tag{12}$$

This is performed for the assessment of position to locate the new elephant and to assess its fitness. Among the solutions, the best and worst ones can be distinguished through the following steps:

Step 5: Rejection of the Worst Solutions

In this step, the worst elephant is separated from their clan and can be understood through the Eq. (13)

$$\beta_{worst,Ci,j} = \beta_{\min} + (\beta_{\max} - L_{\min} + 1) * R \tag{13}$$

where $\beta_{worst,ci,j}$ represents the worst male elephant in clan and $\beta_{\max}$ *and* $\beta_{\min}$ denote the maximum and minimum allowable boundary limits for the clan elephants.

Step 6: Termination process

After ignoring the most exceedingly terrible solutions, the fitness function of the produced rules needs to be assessed. If target condition is optimal (least), then the procedure is finished. In general, stages 2–5 are repeated until the ideal rules are achieved in NS process.

4.4.1 Optimal NS Rules

From the above OEA method, the ideal rules of the data classification process need to be located in an ideal arrangement of fuzzy rules. This is executed by decided a fitting arrangement of NS for quantitative information. To evaluate the accuracy and rule complexity of neutrosophic membership function, fuzzy decision tree is used in this study.

4.5 NS-Based Data Classification Model

In the proposed data classification process, Deep Learning (DL) technique [27–31] is used which functions on the basis of optimal NS rules. This model has various layers of hidden units and the output is named after DNNs. Further, it contains both pre-training and fine-tuning phases during parameter learning stage.

Pre-training phase

In this phase, DBN is employed to design which comprise input, hidden, and output layers. Here, the RBM is executed to overcome the above issue.

Restricted Boltzmann Machine (RBM): RBM is noted as Markov arbitrary field which has one layer of stochastic hidden units and another layer of stochastic visible units.

At first, the visible units' v are to be initiated into training vector.

$$F(v,h) = -\sum_{i=1}^{I}\sum_{j=1}^{J} Opt\_R_{ij}v_i h_j - \sum_{i=1}^{I}\delta_i v_i - \sum_{j=1}^{J}\lambda_j h_j \tag{14}$$

where Opt_R_{ij} represent the symmetric interaction term among the visible u_i and hidden units h_j, δ, λ shows the bias term, and M, N represents the number of visible and hidden units. It is extremely straightforward to get a tolerant instance of $\left(u_i, h_j\right)_{data}$

$$\rho\left(h_j = 1v\right) = \phi\left(\sum_{i=1}^{I} Opt\_R_{ij}v_i + \delta_j\right) \tag{15}$$

The auxiliary of the log probability, for a preparation vector regarding a weight, is unconventionally straightforward. There are no prompt effects found in the hidden units of an RBM. Hidden and visible units are parallel in the offered visible and hidden units. To execute the stochastic steepest ascending in the log probability of arrangement data, there is a way in which a much-simple learning rule is defined and is given in the equation below.

$$\Delta Opt\_R_{ij}\theta\left(v_i h_j\right)_{data}\left(v_i h_j\right)_{reconstruction} \tag{16}$$

While preparing RBM, the unrelated RBM can be loaded over it in order to shape a multilayer model [32]. If the input layer is a trained vector, single RBM is stacked every time and the units are circulated by weights and inclinations.

Fine tuning phase

A typical BackPropagation Algorithm (BPA) works at the time of fine-tuning phase. BPA has 'N' number of input neurons whereas three hidden layers are present in it.

Moreover, the training dataset is trained until the optimized weight is attained with the help of conditions.

Pseudocode for proposed model

Objective: Data classification

Input: Database

Output: Class

Pre Processing: Read the database to remove null and missing attributes

KPCA: Feature Reduction Model

NS Generation

Membership Function & Rule Generation

Rule Optimization using OEHA

Initialize the population

Opposition Process

Fitness Evaluation

Position and Movement updating

Check condition

If satisfied the process end, otherwise check again.

Termination Process

Classification model: Class 0, 1

---

5 Result and Analysis

The proposed optimal NS rule-based data classification model was implemented in MATLAB 2017a with i5 processor and i7 processor. The data, used for analysis, were collected from UCI repository i.e., Hill-Valley, isolet, and musk databases. The proposed model was compared with existing classification and optimization techniques under different performance measures.

Table 1 Attribute information of databases

Database	Number of instances	Number of attributes	Number of web hits	Number of class
Isolet database	7797	617	96,872	5
Musk database	6598	168	54,459	2
Hill-Valley atabase	606	101	59,415	2

5.1 Database Description

Musk database: This dataset is an arrangement of 102 atoms, among which 39 are judged by human specialists to be 'musks' whereas rest of the 63 atoms are judged to be non-musks. The objective is to figure out the prediction method to categorize the new particles as either musks or non-musks. Notwithstanding, it is to be highlighted that the portrayal of these particles relies on correct shape, or adaptation of the atom. Since bonds can pivot, a solitary particle can receive a wide range of shapes. To create this collection of information, all low-vitality adaptations of the particles were created to deliver 6598 compliances. At that point, an element vector was separated that portrayed every compliance.

Isolet Database: This informational index is a result of data collection from 150 subjects who talked the name of each letter twice in an order. Henceforth, there were 52 training cases from every speaker. The speakers were segregated into sets of 30 speakers each and were alluded to isolet1, isolet2, isolet3, isolet4, and isolet5. The information shown in isolet1+2+3+4.data is in consecutive request that first starts with the speakers from isolet1, at that point isolet2, etc. The test set isolet5 is a separate file.

Hill-Valley database: Each record represents 100 points in a two-dimensional graph. When plotted in order (from 1 through 100) as Y co-ordinate, the points create either a hill or a valley, and the information is shown in Table 1.

5.2 Performance Measures

$$\text{Precision: } Pre = \left(\frac{CP}{CP + WP}\right)$$

$$\text{Recall: } Recall = \left(\frac{CP}{CP + WN}\right)$$

$$\text{Accuracy: } Acc = \left(\frac{CP + CN}{CP + CN + WP + WN}\right)$$

$$\text{Specificity: } Spec = \left(\frac{CN}{WP + CN}\right)$$

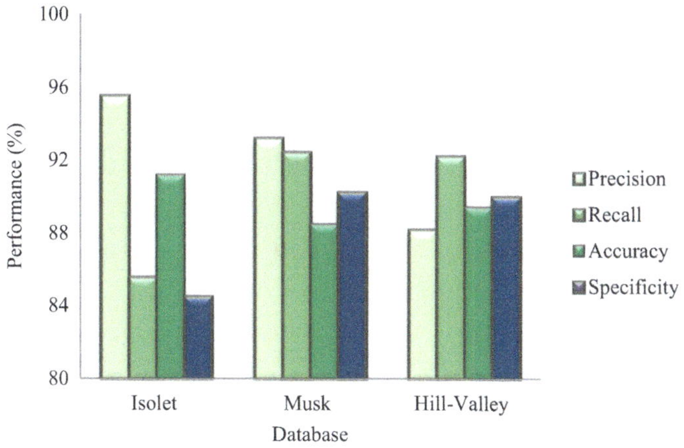

Fig. 4 Results for the proposed model

The results of the proposed classifier for various databases are illustrated in Fig. 4. The evaluated performance metrics were already described in the sections above. For instance, musk database attained a precision of 92.23%, Hill-Valley database achieved 85.56%, and finally, 94.55% precision was achieved in Isolet database by the proposed optimal NS rule-based DNN classifier model. Similarly, a better confusion matrix was attained in the proposed data classification process in other metrics like recall, sensitivity, and specificity. These outcomes were attained from feature reduction (KPCA) and optimal NS set-based classifier. So the number of decision fusion is not necessary to calculate as it is different from combination fusion.

The accuracy analysis of error rate, under various databases, is illustrated in Figs. 5a–c. Here the features were differentiated from 2, 4, 6, and up to 16 feature sets for analysis. The feature subsets, obtained in different runs, were inclusive of same number of features but under different classification error rates. Therefore, the solutions obtained in each run do not dominate whereas some solutions obtained from average OEHA tend to dominate others. An error rate of 13.28% was achieved in Hill-Valley database of the proposed feature selection process. In case of feature being 8, the error rate was 22.25% in isolet database. Similarly, all the features were analyzed and the results are depicted in Fig. 5. Table 2 compares the results of feature selection process. The reduced features are denoted through optimal features' set of each database and technique. The minimum error rate was 0.56 in Musk database with three features in the proposed model i.e., KPCA + NS + OEHA. Similarly, another database (Hill-Valley and isolet) also achieved minimum error rates i.e., 0.49 and 0.77 for the proposed technique compared to KPCA + NS + EHA technique. The minimum difference stood at 10–15% in all the techniques under smaller number of features. At the same time, it also achieved a low classification error rate than all the features. Different classification performance was achieved under distinct, yet

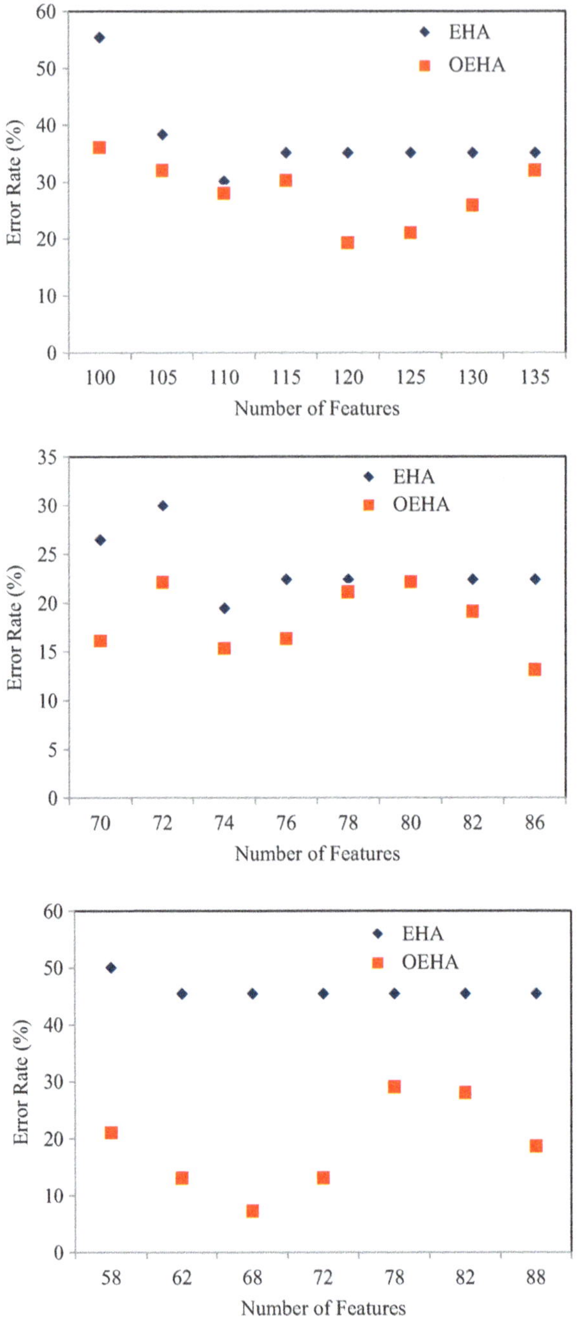

Fig. 5 Feature analysis

Table 2 Proposed feature set analysis of the databases

Technique	Isolet database		Musk database		Hill-Valley database	
	Reduced features	Error rate	Reduced features	Error rate	Reduced features	Error rate
PCA	456	–	109	–	44	–
KPCA	354	–	114	–	94	–
KPCA + NS + GA	256	0.35	102	0.38	68	0.33
KPCA + NS + EHA	135	0.28	86	0.47	88	0.26
KPCA + NS + OEHA	105	0.24	70	0.33	62	0.22

same number of feature combinations. The feature subsets obtained from different runs were inclusive of same number of features while the classification error rates varied.

5.3 *Comparative Analysis*

Figure 6 shows the results of comparative analysis of different databases (Hill-Valley, musk, and heart). Here, Fig. 6a shows the precision results of the proposed classifier NS + OEHA + DL with maximum precision rate i.e., 96.33%, when using Hill-Valley database, 93.33% in musk database and 95.78% in isolet database. When the results were compared with other techniques i.e., NS + EHA + DL and DNN, the precision difference was found to be in the range of 7.56–11.2%. Further, it also reduced the computational time and complexity by removing the overlaps between sets and classes. These results were attained in neutrosophic logic by the introduction of indeterminacy term. Following this, the best rules for classification were framed from hybrid genetic algorithm (from the initial set of rules generated from the considered datasets). Figure 6b shows the results for recall analysis of data classification or sensitivity of the proposed work i.e., 75.56%. When compared with other techniques, the difference was much higher. Likewise, Fig. 6c, d analyzed the results of various databases. Under same number of features in different combinations, different classification performances were achieved. The feature subsets obtained under different runs were inclusive of same number of features under different classification error rates. The significance of each feature was calculated in order to evaluate the importance of the selected feature for classification.

Table 3 demonstrates the testing and training results of data classification process (NS + OEHA + DL). The results were attained using 90% training data with 10% testing data while the execution measures were high when compared to low level

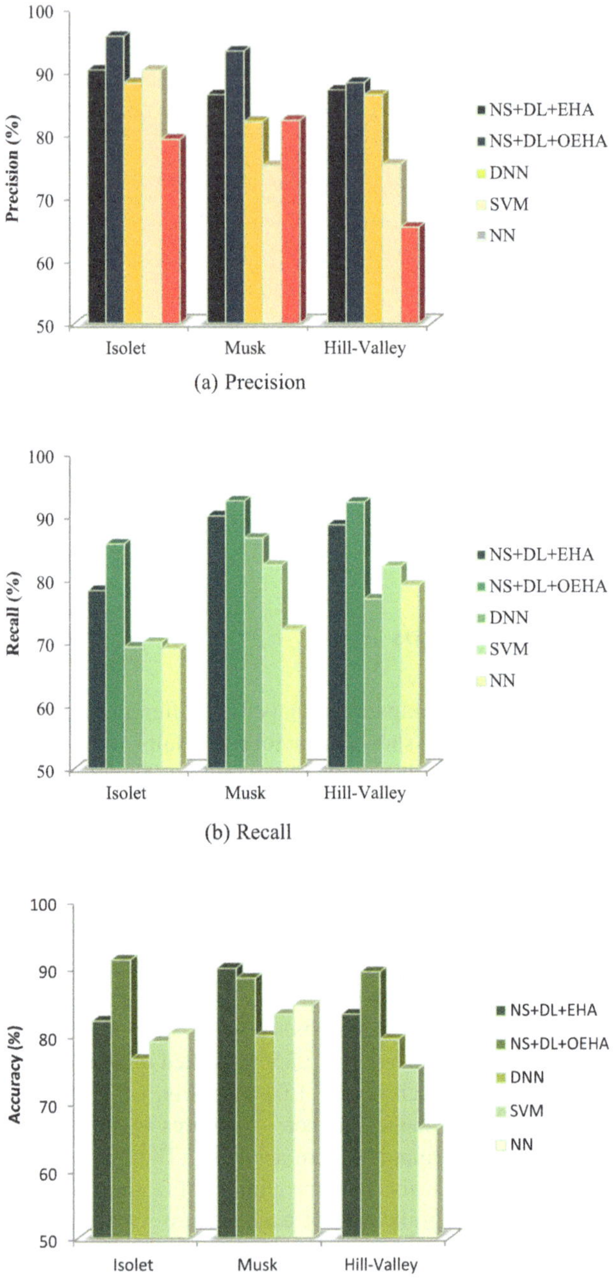

(a) Precision

(b) Recall

(c) Accuracy

Fig. 6 Classification results analysis of different models

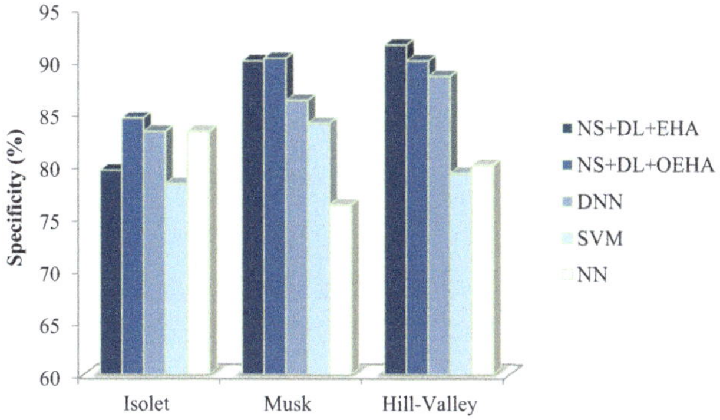

(d) Specificity

Fig. 6 (continued)

training data. For instance, in 80%–20% of training and testing databases, the performance values were 89.52%, 75.22%, and 79.52% for musk and comparable outcomes in databases. This infers that if data is trained at high rates, its accuracy and other performance results can be improved.

The computational time of the classification model is illustrated in Fig. 7. Here various techniques were considered for the investigation such as NS + OEHA + DL, EHA, DNN, NN, and SVM. It is known that all the algorithms can perform one run in moderately brief time i.e., a couple of minutes or even less than one minute. For example, Hill-Valley, isolet, and musk database. In such expansive data sets, the computational time is more compared to small data sets. The minimum computational time was 0.15 min in the proposed classifier and when compared with different systems, it seems to be the minimum. At the time of evolutionary process, the earlier discussed techniques possess the same number of individuals and iterations to evaluate the performance measures.

6 Conclusion

In this paper, the researchers investigated an optimal NS rule-based feature selection model in data classification process. The performance measures of the data classification models and their details were discussed. The major parts of this research study are NS rule generation and rule optimization for which OEHA was introduced. On the basis of OEHA, better rules were framed in a short period of time. Using this classification model, neutrosophic rule-based system attained better results. In addition, the result showed that NS + OHEA + DL is highly robust in classification. These feature selection models improve the quality of feature subset in optimal NS rule framing and optimal granularity parameters. The identification process of relevant

Table 3 Results analysis of proposed model on different dataset

Train-Test (%)	Isolet database			Musk database			Hill-Valley database		
	Precision	Recall	Accuracy	Precision	Recall	Accuracy	Precision	Recall	Accuracy
90–10	95.56	92.22	89.45	93.22	85.56	91.22	86.22	92.42	86.45
80–20	90.22	86.45	76.22	90	82.22	88	85.4	86.45	79.45
70–30	88.22	82.22	72.2	86.52	79.55	82.41	79.22	79.22	76
60–40	83.22	76.45	73	86.45	75	80	75.22	74.4	69.22

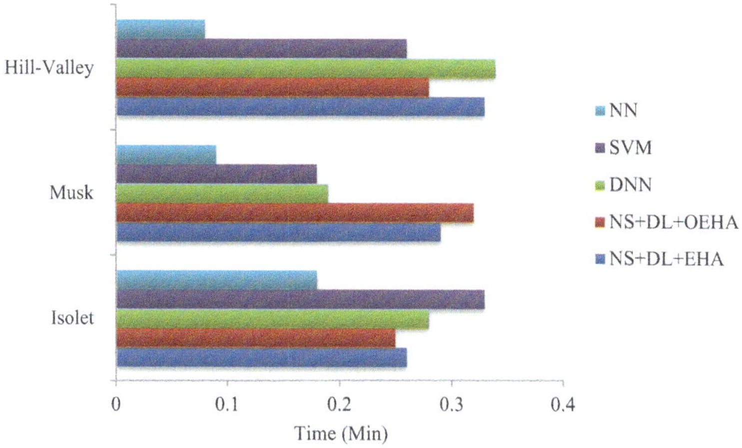

Fig. 7 Computational time analysis of different methods

features is important for mapping the features into a new space of reduced dimensions. One of the challenges is to select the best subset of features from the available original features in the given dimensionality data. From implementation results, the performance measures achieved i.e., precision, recall and specificity were 88.2%, 92.22%, 90% for Hill-Valley data, 95.56%, 85.58% and 84.52% for isolet data and finally 93.22%, 92.45% and 90.2% for musk data. The comparative analysis was also discussed in the above section. In future, the researchers suggest expanding the proposed approach with new measures for medical databases using a hybrid classification technique along with swarm-based feature selection.

References

1. Sheykhizadeh, S., Naseri, A.: An efficient swarm intelligence approach to feature selection based on invasive weed optimization: application to multivariate calibration and classification using spectroscopic data. Spectrochim. Acta Part A Mol. Biomol. Spectros. **194**, 202–210 (2018)
2. Lim, H., Lee, J., Kim, D.W.: Optimization approach for feature selection in multi-label classification. Pattern Recogn. Lett. **89**, 25–30 (2017)
3. Agor, J., Özaltın, O.Y.: Feature selection for classification models via bilevel optimization. Comput. Oper. Res. 1–32 (2018)
4. Brezočnik, L.: Feature selection for classification using particle swarm optimization. In: IEEE EUROCON 2017-17th International Conference on Smart technologies, pp. 966–971. IEEE (2017)
5. Arora, S., Anand, P.: Binary butterfly optimization approaches for feature selection. Expert Syst. Appl. 1–54 (2018)
6. Harde, S., Sahare, V.: Design and implementation of ACO feature selection algorithm for data stream mining. In: International Conference on Automatic Control and Dynamic Optimization Techniques (ICACDOT), pp. 1047–1051. IEEE (2016)

7. Xue, B., Zhang, M., Browne, W.N.: Particle swarm optimization for feature selection in classification: a multi-objective approach. IEEE Trans. Cybern. **43**(6), 1656–1671 (2013)
8. Fong, S., Wong, R., Vasilakos, A.: Accelerated PSO swarm search feature selection for data stream mining big data. IEEE Trans. Serv. Comput. **1**, 1–1 (2016)
9. Ke, W., Wang, Y., Lei, X., Wei, B.: Spark-based feature selection algorithm of network traffic classification. In: 2017 13th International Conference on Computational Intelligence and Security (CIS), pp. 140–144. IEEE (2017)
10. Rani, A.S.S., Rajalaxmi, R.R.: Unsupervised feature selection using binary bat algorithm. In: 2015 2nd International Conference on Electronics and Communication Systems (ICECS), pp. 451–456. IEEE (2015)
11. Hassanien, A.E., Basha, S.H., Abdalla, A.S.: Generalization of fuzzy C-means based on neutrosophic logic. Stud. Inf. Control **27**(1), 43–54 (2018)
12. López-Cabrera, J.D., Lorenzo-Ginori, J.V.: Feature selection for the classification of traced neurons. J. Neurosci. Methods **303**, 41–54 (2018)
13. Kim, K.J., Jun, C.H.: Rough set model based feature selection for mixed-type data with feature space decomposition. Expert Syst. Appl. **103**, 196–205 (2018)
14. Zhang, L., Mistry, K., Lim, C.P., Neoh, S.C.: Feature selection using firefly optimization for classification and regression models. Decis. Support Syst. **106**, 64–85 (2018)
15. Kumar, S.S., Inbarani, H.H.: Optimistic multi-granulation rough set based classification for medical diagnosis. Procedia Comput. Sci. **47**, 374–382 (2015)
16. Ougiaroglou, S., Diamantaras, K.I., Evangelidis, G.: Exploring the effect of data reduction on neural network and support vector machine classification. Neurocomputing **280**, 101–110 (2018)
17. Dong, H., Li, T., Ding, R., Sun, J.: A novel hybrid genetic algorithm with granular information for feature selection and optimization. Appl. Soft Comput. **65**, 33–46 (2018)
18. Sharif, M., Khan, M.A., Iqbal, Z., Azam, M.F., Lali, M.I.U., Javed, M.Y.: Detection and classification of citrus diseases in agriculture based on optimized weighted segmentation and feature selection. Comput. Electron. Agric. **150**, 220–234 (2018)
19. Vinod, D.F., Vasudevan, V.: A filter based feature set selection approach for big data classification of patient records. In: International Conference on Electrical, Electronics, and Optimization Techniques (ICEEOT), pp. 3684–3687. IEEE (2016)
20. Basha, S.H., Abdalla, A.S., Hassanien, A.E.: GNRCS: hybrid classification system based on neutrosophic logic and genetic algorithm. In: 2016 12th International Computer Engineering Conference (ICENCO), pp. 53–58. IEEE (2016)
21. Nalluri, M.S.R., SaiSujana, T., Reddy, K.H., Swaminathan, V.: An efficient feature selection using artificial fish swarm optimization and SVM classifier. In: 2017 International Conference on Networks and Advances in Computational Technologies (NetACT), pp. 407–411. IEEE (2017)
22. Kilany, M., Hassanien, A.E.: A hybrid elephant herding optimization and support vector machines for human behavior identification. In: 2017 Eighth International Conference on Intelligent Computing and Information Systems (ICICIS), pp. 178–184. IEEE (2017)
23. Kamiran, F., Calders, T.: Data preprocessing techniques for classification without discrimination. Knowl. Inf. Syst. **33**(1), 1–33 (2012)
24. Wang, Q.: Kernel principal component analysis and its applications in face recognition and active shape models, pp. 1–9. arXiv:1207.3538 (2012)
25. Kraipeerapun, P., Fung, C.C., Wong, K.W.: Ensemble neural networks using interval neutrosophic sets and bagging. In: Third International Conference on Natural Computation, 2007. ICNC 2007, vol. 1, pp. 386–390. IEEE (2007)
26. Kraipeerapun, P., Fung, C.C., Wong, K.W.: Multiclass classification using neural networks and interval neutrosophic sets, pp. 123–128 (2006)
27. Krishnaraj, N., Elhoseny, M., Thenmozhi, M., Selim, M.M., Shankar, K.: Deep learning model for real-time image compression in Internet of Underwater Things (IoUT). J. Real-Time Image Process. 1–15 (2019)

28. Vinayakumar, R., Soman, K.P., Poornachandran, P., Akarsh, S., Elhoseny, M.: Improved DGA domain names detection and categorization using deep learning architectures with classical machine learning algorithms. In: Cybersecurity and Secure Information Systems, pp. 161–192. Springer, Cham (2019)
29. Vinayakumar, R., Soman, K.P., Poornachandran, P., Akarsh, S., Elhoseny, M.: Deep learning framework for cyber threat situational awareness based on email and URL data analysis. In: Cybersecurity and Secure Information Systems, pp. 87–124. Springer, Cham (2019)
30. Shankar, K., Lakshmanaprabu, S.K., Khanna, A., Tanwar, S., Rodrigues, J.J., Roy, N.R.: Alzheimer detection using group grey wolf optimization based features with convolutional classifier. Comput. Electr. Eng. **77**, 230–243 (2019)
31. Lakshmanaprabu, S.K., Mohanty, S.N., Krishnamoorthy, S., Uthayakumar, J., Shankar, K.: Online clinical decision support system using optimal deep neural networks. Appl. Soft Comput. **81**, 105487 (2019)
32. Yan, Y., Chen, M., Shyu, M.L., Chen, S.C.: Deep learning for imbalanced multimedia data classification. In: 2015 IEEE International Symposium on Multimedia (ISM), pp. 483–488. IEEE (2015)

Fuzzy Driven Optimization Methods and Control Systems

Self-Evolving Interval Type-2 Fuzzy Neural Network Design for the Synchronization of Chaotic Systems

Tien-Loc Le, Chih-Min Lin, Tuan-Tu Huynh, and Sung Kyung Hong

Abstract This study presents a self-evolving interval type-2 fuzzy neural network using function-link network and asymmetric membership function, and it is used to synchronize the chaotic systems. The self-evolving algorithm is applied to autonomously construct the network structure. The function-link network and the asymmetric membership function are applied to enhance the learning capability of the proposed network. The parameters of the proposed network are online updated using the adaptive laws which are obtained by the gradient descent method. Finally, numerical simulation results of chaotic systems synchronization are conducted to illustrate the effectiveness of the proposed control method.

Keywords Self-evolving algorithm · Type-2 fuzzy · Function-link network · Asymmetric membership function · Chaotic systems

1 Introduction

In the past decades, the synchronization of chaotic systems has attracted the attention of many researchers owing to its promising applications in various areas [1]. The synchronization of chaotic systems requires the control signals that force a slave system to follow a master system. Chaotic systems have many specific characteristics, such as highly complex dynamics, inner randomness, self-similarity, strange attractor, and sensitivity to initial conditions [2]. Some notable methods of synchronization chaotic systems are given in [3–15]. In 2017, Mohammadzadeh and Ghaemi

T.-L. Le · S. K. Hong
Faculty of Mechanical and Aerospace, Sejong University, Seoul 143-747(05006), Korea
e-mail: skhong@sejong.ac.kr

T.-L. Le · T.-T. Huynh
Faculty of Mechatronics and Electronics, Lac Hong University, Bien Hoa 810000, Vietnam

C.-M. Lin (✉) · T.-T. Huynh
Department of Electrical Engineering, Yuan Ze University, Taoyuan 32003, Taiwan
e-mail: cml@saturn.yzu.edu.tw

D. Gupta et al. (eds.), *Soft Computing for Data Analytics, Classification Model, and Control*, Studies in Fuzziness and Soft Computing 413,
https://doi.org/10.1007/978-3-030-92026-5_5

provided a type-2 fuzzy cerebellar model articulation controller for synchronization of fractional-order chaotic systems [4]. In 2019, Li and Hernández introduced a novel fuzzy rule-based controllers for synchronization of chaotic systems [7]. In 2020, Wang et al., pro-posed an exponential synchronization of chaotic systems with stochastic perturbations via quantized feed-back control [15].

In the past decades, the neurofuzzy systems (NFSs), which combined the human-like reasoning of fuzzy logic and the learning capabilities of neural networks, have been attracted by many researchers [16, 17]. By having the neural networks connectionist structure, the parameters of NFS can be trained using the back-propagation algorithm. The initial concepts of the fuzzy system were conceived and developed by Zadeh in [18], which can be called type-1 fuzzy logic systems (T1FLS). Since this first work, the T1FLS have been successfully applied in many fields such as control systems, system identification, signal processing, prediction systems, and classification systems, etc. [19–26]. However, it is required the knowledge of experts and experimental data for T1FLS working effectively [27]. The studies in [28, 29] have shown that due to precise sets in membership function, the T1FLS cannot capture well with the uncertainties. These uncertainties can come from the internal system uncertainty of the external disturbances, which are supposed that can be captured well by using type-2 fuzzy logic systems (T2FLS). The T2FLS was first introduced by Zadeh in 1975, which has the uncertainty in membership functions [30]. In order to reduce the complexity of T2FLS, the interval type-2 fuzzy logic systems (IT2FLS) have been developed in [31]. Based on IT2FLS, recently, many fuzzy controllers have been developed and applied widely in various fields [32–43].

Design the network structure for the IT2FLS controller is a time-consuming task, and it affects much to the control performance. Recently, the self-evolving algorithm has been applied to obtain a suitable network structure for IT2FLS [44–48]. In 2017, Chen and Liu provided a self-evolving probabilistic fuzzy neural network for control of a permanent magnet linear synchronous motor [44]. In 2019, Le et al. proposed a self-evolving interval type-2 wavelet cerebellar model articulation control design for uncertain nonlinear systems using PSO [46]. In 2020, Afrakhte and Bayat introduced a self-evolving type-2 fuzzy energy management strategy for multi-microgrid systems [48]. However, most of their controller are complex and re-quest an accurate mathematical system model. By applying the self-evolving algorithm, our proposed controller can automatically generate a new rule or delete unused rules.

Recently, many studies have applied the function-link network (FLN) in their network design to achieve better control performance. The FLN was first introduced by Pao [49]; basically it is a feed-forward neural network, which has a simple structure for fast computation and effectively for function approximation [50]. Some notable studies on FLN were given in [51–57]. The recent studies in [58–60] have shown that by using the asymmetric membership functions (AMFs), the learning capability of the network can be improved, and the fuzzy rules can be optimized. In this study, the design of type-2 asymmetric membership functions (T2AMFs) is constructed by two upper Gaussian membership functions and two lower Gaussian member-ship functions.

According to the above discussions, this study presents a self-evolving interval type-2 fuzzy neural net-work design for the synchronization of chaotic systems using function-link network and asymmetric membership function (SEIT2F-AMF). The main contributions of this study are summarized below: (1) The successful design of the SEIT2F-AMF controller with the adaptive laws for online updating the network parameters; (2) The function-link network and the asymmetric membership function are applied to improve the learning capability and the flexibility of the SEIT2F-AMF controller; (3) The self-evolving algorithm is applied to autonomously construct the network structure; (4) The numerical simulation results of chaotic sys-tem synchronization are given to verify the effectiveness of the proposed method.

The remaining parts of this study are presented as follows. The structure of the SEIT2F-AMF controller is described in Sect. 2. The parameter learning and the structure learning are designed in Sect. 3. The simulation results of chaotic system synchronization are presented in Sect. 4. Finally, the conclusions are given in Sect. 5.

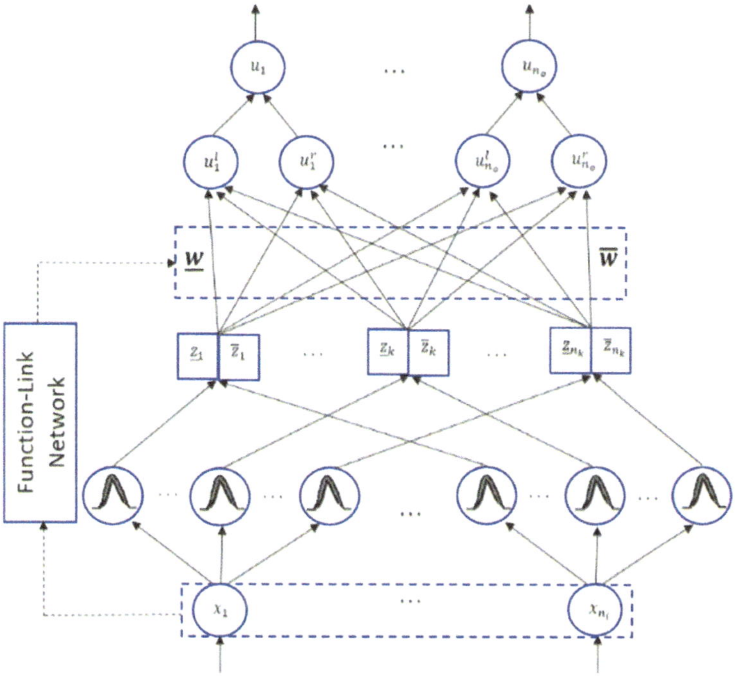

Fig. 1 Structure of the SEIT2F-AMF controller

2 The Structure of the SEIT2F-AMF Controller

The structure of the proposed SEIT2F-AMF controller is depicted in Fig. 1. The process of this fuzzy neural network includes five steps, which are explained in detail as follows.

Step 1: This step is used to accept the input variables, each node in this step corresponds to one input variable.

$$\boldsymbol{X} = \left[x_1, x_2, ..., x_i, ...x_{n_i}\right]^T \quad for \quad i = 1, 2, ..., n_i \tag{1}$$

where n_i is the number of input.

Step 2: The fuzzy membership grade computation is performed in this step. Each node in this step represents a type-2 asymmetric membership function (T2AMF), which is depicted in Fig. 2. The upper and lower membership grade are computed as:

$$\overline{\mu}_{ij} = \begin{cases} \exp\left\{\dfrac{-(x_i - \overline{c}^l_{ij})^2}{2\left(\overline{d}^l_{ij}\right)^2}\right\}, & x_i \le \overline{c}^l_{ij} \\ 1, & \overline{c}^l_{ij} \le x_i \le \overline{c}^r_{ij} \\ \exp\left\{\dfrac{-(x_i - \overline{c}^r_{ij})^2}{2\left(\overline{d}^r_{ij}\right)^2}\right\}, & \overline{c}^r_{ij} \le x_i \end{cases} \quad \text{and}$$

$$\underline{\mu}_{ij} = \begin{cases} \varphi * \exp\left\{\dfrac{-\left(x_i - \underline{c}^l_{ij}\right)^2}{2(\underline{d}^l_{ij})^2}\right\}, & x_i \le \underline{c}^l_{ij} \\ \varphi, & \underline{c}^l_{ij} \le x_i \le \underline{c}^r_{ij} \\ \varphi * \exp\left\{\dfrac{-\left(x_i - \underline{c}^r_{ij}\right)^2}{2(\underline{d}^r_{ij})^2}\right\}, & \underline{c}^r_{ij} \le x_i \end{cases} \tag{2}$$

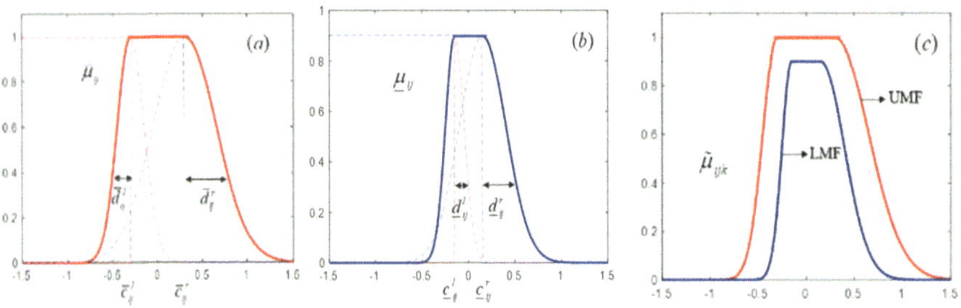

Fig. 2 The type-2 AMF Gaussian membership function: **a** the upper MF, **b** the lower MF, **c** the combined MF

where $j = 1, 2, ..., n_j$ is the number of membership function; $\overline{\mu}_{ij}$ and $\underline{\mu}_{ij}$ are the upper and lower membership functions; $\overline{c}^l_{ij}, \overline{c}^r_{ij}$ and $\underline{c}^l_{ij}, \underline{c}^r_{ij}$ denote the means of the two upper and two lower GMFs; $\overline{d}^l_{ij}, \overline{d}^r_{ij}$ and $\underline{d}^l_{ij}, \underline{d}^r_{ij}$ denote the variances of the two upper and two lower GMFs. To ensure the reasonable of the T2AMF, the following condition is provided

$$\begin{cases} \overline{c}^l_{ij} \le \underline{c}^l_{ij} \le \underline{c}^r_{ij} \le \overline{c}^r_{ij} \\ \underline{d}^l_{ij} \le \overline{d}^l_{ij} \le \underline{d}^r_{ij} \le \overline{d}^r_{ij} \\ 0.5 \le \varphi \le 1 \end{cases} \tag{3}$$

Step 3: The fuzzy firing strengths are calculated in this step based on product t-norm operator.

$$\overline{z}_k = \prod_{j=1}^{n_j} \overline{\mu}_{ij} \quad \text{and} \quad \underline{z}_k = \prod_{j=1}^{n_j} \underline{\mu}_{ij} \tag{4}$$

where $\overline{z}_k$ and $\underline{z}_k$ are the upper and lower firing strength of the kth rule.

Step 4: This step computes the fuzzy weights based on the current inputs and the FLN. The structure of the proposed FLN is depicted in Fig. 3.

Firstly, the trigonometric function is used to expand the inputs, and it is shown as:

$$\mathbf{\Psi} = \left[i_1, \sin(\pi i_1), \cos(\pi i_1), ..., i_{n_i}, \sin\left(\pi i_{n_i}\right), \cos\left(\pi i_{n_i}\right), i_1 i_2, ..., i_1 i_{n_i}\right]^T \tag{5}$$

where $\mathbf{\Psi} = \left[\psi_1, \psi_2 ..., \psi_{n_m}\right]^T$ is the output function expansion vector, $m = 1, 2, ..., n_m$ is the number output of function expansion.

Fig. 3 Structure of the FLN

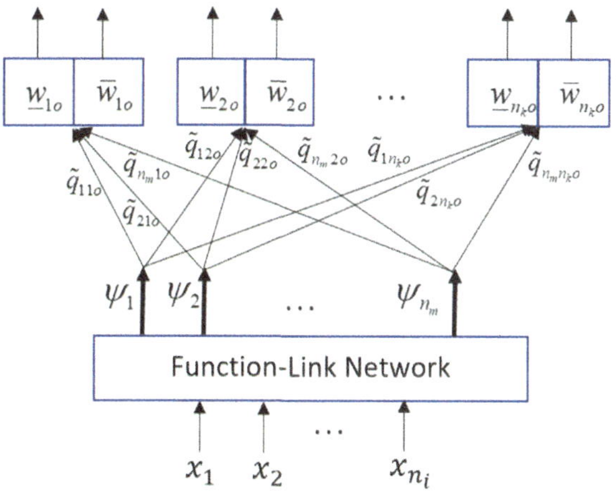

Then, the output of FLN is computed by the output function expansion vector and the functional connecting weights

$$\begin{bmatrix} \overline{w}_{1o} \\ \vdots \\ \overline{w}_{n_k o} \end{bmatrix} = \begin{bmatrix} \overline{q}_{11o}, \overline{q}_{21o} & \cdots & \overline{q}_{n_m 1o} \\ \vdots & \ddots & \vdots \\ \overline{q}_{1n_k o}, \overline{q}_{2n_k o} & \cdots & \overline{q}_{n_m n_k o} \end{bmatrix} \begin{bmatrix} \psi_1 \\ \vdots \\ \psi_{n_m} \end{bmatrix} = \overline{\boldsymbol{q}}_1^T \boldsymbol{\psi} \quad \text{and}$$

$$\begin{bmatrix} \underline{w}_{1o} \\ \vdots \\ \underline{w}_{n_k o} \end{bmatrix} = \begin{bmatrix} \underline{q}_{11o}, \underline{q}_{21o} & \cdots & \underline{q}_{n_m 1o} \\ \vdots & \ddots & \vdots \\ \underline{q}_{1n_k o}, \underline{q}_{2n_k o} & \cdots & \underline{q}_{n_m n_k o} \end{bmatrix} \begin{bmatrix} \psi_1 \\ \vdots \\ \psi_{n_m} \end{bmatrix} = \underline{\boldsymbol{q}}_1^T \boldsymbol{\psi} \tag{6}$$

where $\overline{\boldsymbol{q}}_1$ and $\underline{\boldsymbol{q}}_1$ are the upper and lower functional connecting weights vectors; $\overline{w}_{ko}$ and $\underline{w}_{ko}$ are the upper and lower fuzzy weights for the kth rule, and oth output.

Step 5: This step performs the defuzzification using center of gravity method. The Karnik–Mendel (KM) algorithm is applied to compute type reduction of interval type-2 fuzzy sets. The output of this step is obtained as:

$$\hat{u}^o_{SEIT2F-AMF} = \frac{u^l_o + u^r_o}{2} = \frac{\sum_{k=1}^{n_k} z^l_k \underline{w}_{ko}}{\sum_{k=1}^{n_k} z^l_k} + \frac{\sum_{k=1}^{n_k} z^r_k \overline{w}_{ko}}{\sum_{k=1}^{n_k} z^r_k} \tag{7}$$

where f^i_l and f^i_r are the firing strengths, which are computed as

$$z^l_k = \begin{cases} \overline{z}_k, & k \le L \\ \underline{z}_k, & k > L \end{cases} \quad \text{and} \quad z^r_k = \begin{cases} \underline{z}_k, & k \le R \\ \overline{z}_k, & k > R \end{cases} \tag{8}$$

where L and R denote the left and right switch points, respectively, which can be computed using KM algorithm [61].

3 Parameter Learning and Structure Learning Algorithms

3.1 Parameter Learning Algorithm

Consider the following nonlinear system:

$$\begin{aligned} \boldsymbol{x}^{(n)}(t) &= \boldsymbol{F}(\boldsymbol{x}(t)) + \boldsymbol{G}(\boldsymbol{x}(t))\boldsymbol{u}(t) + \boldsymbol{v}(t) \\ &= \boldsymbol{F}_0(\boldsymbol{x}(t)) + \Delta\boldsymbol{F}(\boldsymbol{x}(t)) + [\boldsymbol{G}_0(\boldsymbol{x}(t)) + \Delta\boldsymbol{G}(\boldsymbol{x}(t))]\boldsymbol{u}(t) + \boldsymbol{v}(t) \\ &= \boldsymbol{F}_0(\boldsymbol{x}(t)) + \boldsymbol{G}_0(\boldsymbol{x}(t))\boldsymbol{u}(t) + \boldsymbol{\zeta}(t) \end{aligned} \tag{9}$$

where $\boldsymbol{x}(t)$ and $\boldsymbol{u}(t)$ denote the system output vector and the control input vector, respectively; $\boldsymbol{F}(\boldsymbol{x})$ and $\boldsymbol{G}(x)$ denote the bounded nonlinear functions; $\boldsymbol{F}_0(\boldsymbol{x}(t))$, $\boldsymbol{G}_0(\boldsymbol{x}(t))$ and $\Delta\boldsymbol{F}(\boldsymbol{x}(t))$, $\Delta\boldsymbol{G}(\boldsymbol{x}(t))$ denote the nominal terms and the unknown uncertainties terms for $\boldsymbol{F}(\boldsymbol{x})$ and $\boldsymbol{G}(\boldsymbol{x})$, respectively; assume $\boldsymbol{G}_0(\boldsymbol{x}) = diag[g_{0i}]$, $i = 1, 2, ..., n$; $\boldsymbol{v}(t)$ denotes the external disturbance; $\boldsymbol{\zeta}(t)$ denotes the lumped uncertainty term, which is given by

$$\boldsymbol{\zeta}(t) = \Delta\boldsymbol{F}(\boldsymbol{x}(t)) + \Delta\boldsymbol{G}(\boldsymbol{x}(t))\boldsymbol{u}(t) + \boldsymbol{v}(t) \tag{10}$$

Assume $\boldsymbol{\zeta}(t)$ is bounded by a bound $\boldsymbol{C}(t)$, such that $|\boldsymbol{\zeta}(t)| \leq C(t)$.

The system tracking error vector is defined as

$$\boldsymbol{E}(t) = \left[\boldsymbol{e}^T(t), \dot{\boldsymbol{e}}^T(t), ..., \ \boldsymbol{e}^{T(n-1)}(t)\right]^T \tag{11}$$

where

$$\boldsymbol{e}(t) = \boldsymbol{x}_{ref}(t) - \boldsymbol{x}(t) \tag{12}$$

where $\boldsymbol{e}(t)$ and $\boldsymbol{x}_{ref}(t)$ denote the tracking error and the desired signal, respectively.

From (9), if $\boldsymbol{F}_0(\boldsymbol{x}(t))$, $\boldsymbol{G}_0(\boldsymbol{x}(t))$ and $\boldsymbol{\zeta}(t)$ are exactly known, an ideal controller can be designed as:

$$\boldsymbol{u}^*(t) = \boldsymbol{G}_0^{-1}(\boldsymbol{x}(t))\left[\boldsymbol{x}_{ref}^{(n)}(t) - \boldsymbol{F}_0(\boldsymbol{x}(t)) - \boldsymbol{\zeta}(t) + \boldsymbol{K}^T\boldsymbol{E}(t)\right] \tag{13}$$

where $\boldsymbol{K}^T$ is the feedback gain matrix.

Substituting (13) into (9), obtain

$$\boldsymbol{e}^{(n)} + \boldsymbol{K}^T\boldsymbol{e} = 0 \tag{14}$$

In (14), if $\boldsymbol{K}$ is selected for satisfying the Hurwitz stability, then $\lim_{t\to\infty} \boldsymbol{e}(t) = 0$. However, the terms $\boldsymbol{F}_0(\boldsymbol{x}(t))$, $\boldsymbol{G}_0(\boldsymbol{x}(t))$ and $\boldsymbol{\zeta}(t)$ are unknown. Therefore, the ideal controller $\boldsymbol{u}^*(t)$ is unobtainable. Hence, in this study, an optimal SEIT2F-AMF controller is proposed to approximate $\boldsymbol{u}^*(t)$ as:

$$\boldsymbol{u}^*(t) = \boldsymbol{u}^*_{SEIT2F-AMF}(\underline{q}^*, \overline{q}^*, \underline{c}^{l*}, \overline{c}^{l*}, \underline{c}^{r*}, \overline{c}^{r*}, \underline{d}^{l*}, \overline{d}^{l*}, \underline{d}^{r*}, \overline{d}^{r*}, t) + \boldsymbol{\tau}(t) \tag{15}$$

where $\underline{q}^*, \overline{q}^*, \underline{c}^{l*}, \overline{c}^{l*}, \underline{c}^{r*}, \overline{c}^{r*}, \underline{d}^{l*}, \overline{d}^{l*}, \underline{d}^{r*}, \overline{d}^{r*}$ are the optimal parameters for $\underline{q}, \overline{q}, \underline{c}^l, \overline{c}^l, \underline{c}^r, \overline{c}^r, \underline{d}^l, \overline{d}^l, \underline{d}^r, \overline{d}^r$; $\boldsymbol{\tau}(t)$ is the approximation error.

Since the $\boldsymbol{u}^*_{SEIT2F-AMF}$ cannot be achieved, an estimation controller, $\hat{\boldsymbol{u}}_{SEIT2F-AMF}$, is designed to estimate $\boldsymbol{u}^*_{SEIT2F-AMF}$. Then the control signal can be designed as

$$\hat{\boldsymbol{u}}(t) = \hat{\boldsymbol{u}}_{SEIT2F-AMF}(\hat{\underline{q}}, \hat{\overline{q}}, \hat{\underline{c}}^l, \hat{\overline{c}}^l, \hat{\underline{c}}^r, \hat{\overline{c}}^r, \hat{\underline{d}}^l, \hat{\overline{d}}^l, \hat{\underline{d}}^r, \hat{\overline{d}}^r, t) + \hat{\boldsymbol{u}}_C(t) \tag{16}$$

where $\hat{\underline{q}}, \hat{\overline{q}}, \hat{\underline{c}}^l, \hat{\overline{c}}^l, \hat{\underline{c}}^r, \hat{\overline{c}}^r, \hat{\underline{d}}^l, \hat{\overline{d}}^l, \hat{\underline{d}}^r, \hat{\overline{d}}^r$ are the estimation of $\underline{q}^*, \overline{q}^*, \underline{c}^{l*}, \overline{c}^{l*}, \underline{c}^{r*}, \overline{c}^{r*}, \underline{d}^{l*}, \overline{d}^{l*}, \underline{d}^{r*}, \overline{d}^{r*}$; $\hat{\boldsymbol{u}}_C$ is the robust compensator controller, which is given as

$$\hat{\boldsymbol{u}}_C(t) = \hat{\boldsymbol{C}}(t)\mathrm{sgn}(\boldsymbol{s}(t)) \tag{17}$$

where $\hat{\boldsymbol{C}}(t)$ is the estimated value of the uncertainty bound $\boldsymbol{C}(t)$; $s(t)$ denotes the high-order sliding surface, which is defined as:

$$\boldsymbol{s}(t) = \boldsymbol{e}^{(n-1)} + \boldsymbol{k}_1\boldsymbol{e}^{(n-2)} \ldots + \boldsymbol{k}_n \int_0^t \boldsymbol{e}(\tau)d\tau \tag{18}$$

Let $\boldsymbol{s}(t) = [s_1(t), s_2(t), ..., s_n(t)]^T$. Consider a Lyapunov function as follows:

$$V_1(\boldsymbol{s}(t)) = \frac{1}{2}\boldsymbol{s}^T(t)\boldsymbol{s}(t) \tag{19}$$

Taking time-derivative of (18), one has

$$\dot{\boldsymbol{s}}(t) = \boldsymbol{e}^{(n)} + \boldsymbol{K}^T\boldsymbol{E} \tag{20}$$

where $\boldsymbol{K} = [\boldsymbol{k}_n, ..., \boldsymbol{k}_2, \boldsymbol{k}_1]^T$ is the positive gain matrix, where $\boldsymbol{k}_i = diag(k_{i1}, k_{i2}, ..., k_{in})$, and k_{ij} is also a positive gain for $j = 1, 2, ..., n$.

Taking time-derivative of (19) and using (9), (16), (20), one has

$$\begin{aligned}\dot{V}_1(\boldsymbol{s}(t)) &= \boldsymbol{s}^T(t)\dot{\boldsymbol{s}}(t) = \boldsymbol{s}^T(t)\left[\boldsymbol{e}^{(n)} + \boldsymbol{K}^T\boldsymbol{E}\right] \\ &= \boldsymbol{s}^T(t)\Big[\boldsymbol{x}_d^{(n)}(t) - (\boldsymbol{F}_0(\boldsymbol{x}(t)) + \boldsymbol{G}_0(\boldsymbol{x}(t)) \\ &\quad \left(\hat{\boldsymbol{u}}_{SEIT2F-AMF} + \hat{\boldsymbol{u}}_C(t)\right) + \boldsymbol{\xi}(t)\big) + \boldsymbol{K}^T\boldsymbol{E}\Big]\end{aligned} \tag{21}$$

Applying the gradient descent method and chain rule, the parameters update laws for $\hat{\boldsymbol{u}}_{SEIT2F-AMF}$ are given as

$$\begin{aligned}\hat{\overline{q}}_{mko}(t+1) &= \hat{\overline{q}}_{mko}(t) - \hat{\eta}_q \frac{\partial \dot{V}_1(t)}{\partial \hat{\overline{q}}_{mko}} = \hat{\overline{q}}_{mko}(t) \\ &\quad - \hat{\eta}_q \frac{\partial \dot{V}_1(t)}{\partial \hat{u}^o_{SEIT2F-AMF}} \frac{\partial \hat{u}^o_{SEIT2F-AMF}}{\partial u^r_o} \frac{\partial u^r_o}{\partial \hat{\overline{w}}_{ko}} \frac{\partial \hat{\overline{w}}_{ko}}{\partial \hat{\overline{q}}_{mko}} \\ &= \hat{\overline{q}}_{mko}(t) + \frac{1}{2}\hat{\eta}_q g_{0k} s_k(t) \frac{z^r_k}{\sum_{k=1}^{n_k} z^r_k}\psi_m\end{aligned} \tag{22}$$

$$
\begin{aligned}
\hat{\underline{q}}_{mko}(t+1) &= \hat{\underline{q}}_{mko}(t) - \hat{\eta}_q \frac{\partial \dot{V}_1(t)}{\partial \hat{\underline{q}}_{mko}} = \hat{\underline{q}}_{mko}(t) \\
&\quad - \hat{\eta}_q \frac{\partial \dot{V}_1(t)}{\partial \hat{u}^o_{SEIT2F-AMF}} \frac{\partial \hat{u}^o_{SEIT2F-AMF}}{\partial u^l_o} \frac{\partial u^l_o}{\partial \hat{\underline{w}}_{ko}} \frac{\partial \hat{\underline{w}}_{ko}}{\partial \hat{\underline{q}}_{mko}} \\
&= \hat{\underline{q}}_{mko}(t) + \frac{1}{2}\hat{\eta}_q g_{0k} s_k(t) \frac{z^l_k}{\sum_{k=1}^{n_k} z^l_k} \psi_m
\end{aligned} \tag{23}
$$

$$
\begin{aligned}
\hat{\underline{c}}^l_{ij}(t+1) &= \hat{\underline{c}}^l_{ij}(t) - \hat{\eta}_c \frac{\partial \dot{V}_1(t)}{\partial \hat{\underline{c}}^l_{ij}} = \hat{\underline{c}}^l_{ij}(t) - \hat{\eta}_c \frac{\partial \dot{V}_1(t)}{\partial \hat{u}^o_{SEIT2F-AMF}} \\
&\quad \left(\frac{\partial \hat{u}^o_{SEIT2F-AMF}}{\partial u^l_o} \frac{\partial u^l_o}{\partial z^l_k} \frac{\partial z^l_k}{\partial \hat{\underline{c}}^l_{ij}} + \frac{\partial \hat{u}^o_{SEIT2F-AMF}}{\partial u^r_o} \frac{\partial u^r_o}{\partial z^r_k} \frac{\partial z^r_k}{\partial \hat{\underline{c}}^l_{ij}} \right) \\
&= \hat{\underline{c}}^l_{ij}(t) + \frac{1}{2}\hat{\eta}_c g_{0k} s_k(t) \left(\frac{(\underline{w}_{ko} - u^l_o)}{\sum_{k=1}^{n_k} z^l_k} \frac{\partial z^l_k}{\partial \hat{\underline{c}}^l_{ij}} + \frac{(\overline{w}_{ko} - u^r_o)}{\sum_{k=1}^{n_k} z^r_k} \frac{\partial z^r_k}{\partial \hat{\underline{c}}^l_{ij}} \right)
\end{aligned} \tag{24}
$$

$$
\begin{aligned}
\hat{\underline{c}}^r_{ij}(t+1) &= \hat{\underline{c}}^r_{ij}(t) - \hat{\eta}_c \frac{\partial \dot{V}_1(t)}{\partial \hat{\underline{c}}^r_{ij}} = \hat{\underline{c}}^r_{ij}(t) - \hat{\eta}_c \frac{\partial \dot{V}_1(t)}{\partial \hat{u}^o_{SEIT2F-AMF}} \\
&\quad \left(\frac{\partial \hat{u}^o_{SEIT2F-AMF}}{\partial u^l_o} \frac{\partial u^l_o}{\partial z^l_k} \frac{\partial z^l_k}{\partial \hat{\underline{c}}^r_{ij}} + \frac{\partial \hat{u}^o_{SEIT2F-AMF}}{\partial u^r_o} \frac{\partial u^r_o}{\partial z^r_k} \frac{\partial z^r_k}{\partial \hat{\underline{c}}^r_{ij}} \right) \\
&= \hat{\underline{c}}^r_{ij}(t) + \frac{1}{2}\hat{\eta}_c g_{0k} s_k(t) \left(\frac{(\underline{w}_{ko} - u^l_o)}{\sum_{k=1}^{n_k} z^l_k} \frac{\partial z^l_k}{\partial \hat{\underline{c}}^r_{ij}} + \frac{(\overline{w}_{ko} - u^r_o)}{\sum_{k=1}^{n_k} z^r_k} \frac{\partial z^r_k}{\partial \hat{\underline{c}}^r_{ij}} \right)
\end{aligned} \tag{25}
$$

$$
\begin{aligned}
\hat{\underline{d}}^l_{ij}(t+1) &= \hat{\underline{d}}^l_{ij}(t) - \hat{\eta}_d \frac{\partial \dot{V}_1(t)}{\partial \hat{\underline{d}}^l_{ij}} = \hat{\underline{d}}^l_{ij}(t) - \hat{\eta}_d \frac{\partial \dot{V}_1(t)}{\partial \hat{u}^o_{SEIT2F-AMF}} \\
&\quad \left(\frac{\partial \hat{u}^o_{SEIT2F-AMF}}{\partial u^l_o} \frac{\partial u^l_o}{\partial z^l_k} \frac{\partial z^l_k}{\partial \hat{\underline{d}}^l_{ij}} + \frac{\partial \hat{u}^o_{SEIT2F-AMF}}{\partial u^r_o} \frac{\partial u^r_o}{\partial z^r_k} \frac{\partial z^r_k}{\partial \hat{\underline{d}}^l_{ij}} \right) \\
&= \hat{\underline{d}}^l_{ij}(t) + \frac{1}{2}\hat{\eta}_d g_{0k} s_k(t) \left(\frac{(\underline{w}_{ko} - u^l_o)}{\sum_{k=1}^{n_k} z^l_k} \frac{\partial z^l_k}{\partial \hat{\underline{d}}^l_{ij}} + \frac{(\overline{w}_{ko} - u^r_o)}{\sum_{k=1}^{n_k} z^r_k} \frac{\partial z^r_k}{\partial \hat{\underline{d}}^l_{ij}} \right)
\end{aligned} \tag{26}
$$

$$
\begin{aligned}
\hat{\underline{d}}^r_{ij}(t+1) &= \hat{\underline{d}}^r_{ij}(t) - \hat{\eta}_d \frac{\partial \dot{V}_1(t)}{\partial \hat{\underline{d}}^r_{ij}} = \hat{\underline{d}}^r_{ij}(t) - \hat{\eta}_d \frac{\partial \dot{V}_1(t)}{\partial \hat{u}^o_{SEIT2F-AMF}} \\
&\quad \left(\frac{\partial \hat{u}^o_{SEIT2F-AMF}}{\partial u^l_o} \frac{\partial u^l_o}{\partial z^l_k} \frac{\partial z^l_k}{\partial \hat{\underline{d}}^r_{ij}} + \frac{\partial \hat{u}^o_{SEIT2F-AMF}}{\partial u^r_o} \frac{\partial u^r_o}{\partial z^r_k} \frac{\partial z^r_k}{\partial \hat{\underline{d}}^r_{ij}} \right) \\
&= \hat{\underline{d}}^r_{ij}(t) + \frac{1}{2}\hat{\eta}_d g_{0k} s_k(t) \left(\frac{(\underline{w}_{ko} - u^l_o)}{\sum_{k=1}^{n_k} z^l_k} \frac{\partial z^l_k}{\partial \hat{\underline{d}}^r_{ij}} + \frac{(\overline{w}_{ko} - u^r_o)}{\sum_{k=1}^{n_k} z^r_k} \frac{\partial z^r_k}{\partial \hat{\underline{d}}^r_{ij}} \right)
\end{aligned} \tag{27}
$$

$$\hat{\bar{c}}^{l}_{ij}(t+1) = \hat{\bar{c}}^{l}_{ij}(t) - \hat{\eta}_c \frac{\partial \dot{V}_1(t)}{\partial \hat{\bar{c}}^{l}_{ij}} = \hat{\bar{c}}^{l}_{ij}(t) - \hat{\eta}_c \frac{\partial \dot{V}_1(t)}{\partial \hat{u}^{o}_{SEIT2F-AMF}}$$

$$\left(\frac{\partial \hat{u}^{o}_{SEIT2F-AMF}}{\partial u^{l}_{o}} \frac{\partial u^{l}_{o}}{\partial z^{l}_{k}} \frac{\partial z^{l}_{k}}{\partial \hat{\bar{c}}^{l}_{ij}} + \frac{\partial \hat{u}^{o}_{SEIT2F-AMF}}{\partial u^{r}_{o}} \frac{\partial u^{r}_{o}}{\partial z^{r}_{k}} \frac{\partial z^{r}_{k}}{\partial \hat{\bar{c}}^{l}_{ij}} \right)$$

$$= \hat{\bar{c}}^{l}_{ij}(t) + \frac{1}{2}\hat{\eta}_c g_{0k} s_k(t) \left(\frac{\left(\underline{w}_{ko} - u^{l}_{o}\right)}{\sum_{k=1}^{n_k} z^{l}_{k}} \frac{\partial z^{l}_{k}}{\partial \hat{\bar{c}}^{l}_{ij}} + \frac{\left(\overline{w}_{ko} - u^{r}_{o}\right)}{\sum_{k=1}^{n_k} z^{r}_{k}} \frac{\partial z^{r}_{k}}{\partial \hat{\bar{c}}^{l}_{ij}} \right) \tag{28}$$

$$\hat{\bar{c}}^{r}_{ij}(t+1) = \hat{\bar{c}}^{r}_{ij}(t) - \hat{\eta}_c \frac{\partial \dot{V}_1(t)}{\partial \hat{\bar{c}}^{r}_{ij}} = \hat{\bar{c}}^{r}_{ij}(t) - \hat{\eta}_c \frac{\partial \dot{V}_1(t)}{\partial \hat{u}^{o}_{SEIT2F-AMF}}$$

$$\left(\frac{\partial \hat{u}^{o}_{SEIT2F-AMF}}{\partial u^{l}_{o}} \frac{\partial u^{l}_{o}}{\partial z^{l}_{k}} \frac{\partial z^{l}_{k}}{\partial \hat{\bar{c}}^{r}_{ij}} + \frac{\partial \hat{u}^{o}_{SEIT2F-AMF}}{\partial u^{r}_{o}} \frac{\partial u^{r}_{o}}{\partial z^{r}_{k}} \frac{\partial z^{r}_{k}}{\partial \hat{\bar{c}}^{r}_{ij}} \right)$$

$$= \hat{\bar{c}}^{r}_{ij}(t) + \frac{1}{2}\hat{\eta}_c g_{0k} s_k(t) \left(\frac{\left(\underline{w}_{ko} - u^{l}_{o}\right)}{\sum_{k=1}^{n_k} z^{l}_{k}} \frac{\partial z^{l}_{k}}{\partial \hat{\bar{c}}^{r}_{ij}} + \frac{\left(\overline{w}_{ko} - u^{r}_{o}\right)}{\sum_{k=1}^{n_k} z^{r}_{k}} \frac{\partial z^{r}_{k}}{\partial \hat{\bar{c}}^{r}_{ij}} \right) \tag{29}$$

$$\hat{\bar{d}}^{l}_{ij}(t+1) = \hat{\bar{d}}^{l}_{ij}(t) - \hat{\eta}_d \frac{\partial \dot{V}_1(t)}{\partial \hat{\bar{d}}^{l}_{ij}} = \hat{\bar{d}}^{l}_{ij}(t) - \hat{\eta}_d \frac{\partial \dot{V}_1(t)}{\partial \hat{u}^{o}_{SEIT2F-AMF}}$$

$$\left(\frac{\partial \hat{u}^{o}_{SEIT2F-AMF}}{\partial u^{l}_{o}} \frac{\partial u^{l}_{o}}{\partial z^{l}_{k}} \frac{\partial z^{l}_{k}}{\partial \hat{\bar{d}}^{l}_{ij}} + \frac{\partial \hat{u}^{o}_{SEIT2F-AMF}}{\partial u^{r}_{o}} \frac{\partial u^{r}_{o}}{\partial z^{r}_{k}} \frac{\partial z^{r}_{k}}{\partial \hat{\bar{d}}^{l}_{ij}} \right)$$

$$= \hat{\bar{d}}^{l}_{ij}(t) + \frac{1}{2}\hat{\eta}_d g_{0k} s_k(t) \left(\frac{\left(\underline{w}_{ko} - u^{l}_{o}\right)}{\sum_{k=1}^{n_k} z^{l}_{k}} \frac{\partial z^{l}_{k}}{\partial \hat{\bar{d}}^{l}_{ij}} + \frac{\left(\overline{w}_{ko} - u^{r}_{o}\right)}{\sum_{k=1}^{n_k} z^{r}_{k}} \frac{\partial z^{r}_{k}}{\partial \hat{\bar{d}}^{l}_{ij}} \right) \tag{30}$$

$$\hat{\bar{d}}^{r}_{ij}(t+1) = \hat{\bar{d}}^{r}_{ij}(t) - \hat{\eta}_d \frac{\partial \dot{V}_1(t)}{\partial \hat{\bar{d}}^{r}_{ij}} = \hat{\bar{d}}^{r}_{ij}(t) - \hat{\eta}_d \frac{\partial \dot{V}_1(t)}{\partial \hat{u}^{o}_{SEIT2F-AMF}}$$

$$\left(\frac{\partial \hat{u}^{o}_{SEIT2F-AMF}}{\partial u^{l}_{o}} \frac{\partial u^{l}_{o}}{\partial z^{l}_{k}} \frac{\partial z^{l}_{k}}{\partial \hat{\bar{d}}^{r}_{ij}} + \frac{\partial \hat{u}^{o}_{SEIT2F-AMF}}{\partial u^{r}_{o}} \frac{\partial u^{r}_{o}}{\partial z^{r}_{k}} \frac{\partial z^{r}_{k}}{\partial \hat{\bar{d}}^{l}_{ij}} \right)$$

$$= \hat{\bar{d}}^{r}_{ij}(t) + \frac{1}{2}\hat{\eta}_d g_{0k} s_k(t) \left(\frac{\left(\underline{w}_{ko} - u^{l}_{o}\right)}{\sum_{k=1}^{n_k} z^{l}_{k}} \frac{\partial z^{l}_{k}}{\partial \hat{\bar{d}}^{r}_{ij}} + \frac{\left(\overline{w}_{ko} - u^{r}_{o}\right)}{\sum_{k=1}^{n_k} z^{r}_{k}} \frac{\partial z^{r}_{k}}{\partial \hat{\bar{d}}^{r}_{ij}} \right) \tag{31}$$

where $\hat{\eta}_q$, $\hat{\eta}_c$, $\hat{\eta}_d$ are the positive learning-rates.

The derivative terms of z^{l}_{k} and z^{r}_{k} in (24)–(31) can be $\underline{z}_k$ or $\overline{z}_k$, considering both cases as follows:

$$\frac{\partial \bar{z}_k}{\partial \hat{\underline{c}}^l_{ij}} = 0; \quad \frac{\partial \underline{z}_k}{\partial \hat{\underline{c}}^l_{ij}} = \frac{\partial \underline{z}_k}{\partial \underline{\mu}_{ij}} \frac{\partial \underline{\mu}_{ij}}{\partial \hat{\underline{c}}^l_{ij}} = \frac{\underline{z}_k}{\underline{\mu}_{ij}} \frac{\partial \underline{\mu}_{ij}}{\partial \hat{\underline{c}}^l_{ij}}; \tag{32}$$

$$\frac{\partial \bar{z}_k}{\partial \hat{\underline{c}}^r_{ij}} = 0; \quad \frac{\partial \underline{z}_k}{\partial \hat{\underline{c}}^r_{ij}} = \frac{\partial \underline{z}_k}{\partial \underline{\mu}_{ij}} \frac{\partial \underline{\mu}_{ij}}{\partial \hat{\underline{c}}^r_{ij}} = \frac{\underline{z}_k}{\underline{\mu}_{ij}} \frac{\partial \underline{\mu}_{ij}}{\partial \hat{\underline{c}}^r_{ij}}; \tag{33}$$

$$\frac{\partial \bar{z}_k}{\partial \hat{\underline{d}}^l_{ij}} = 0; \quad \frac{\partial \underline{z}_k}{\partial \hat{\underline{d}}^l_{ij}} = \frac{\partial \underline{z}_k}{\partial \underline{\mu}_{ij}} \frac{\partial \underline{\mu}_{ij}}{\partial \hat{\underline{d}}^l_{ij}} = \frac{\underline{z}_k}{\underline{\mu}_{ij}} \frac{\partial \underline{\mu}_{ij}}{\partial \hat{\underline{d}}^l_{ij}}; \tag{34}$$

$$\frac{\partial \bar{z}_k}{\partial \hat{\underline{d}}^r_{ij}} = 0; \quad \frac{\partial \underline{z}_k}{\partial \hat{\underline{d}}^r_{ij}} = \frac{\partial \underline{z}_k}{\partial \underline{\mu}_{ij}} \frac{\partial \underline{\mu}_{ij}}{\partial \hat{\underline{d}}^r_{ij}} = \frac{\underline{z}_k}{\underline{\mu}_{ij}} \frac{\partial \underline{\mu}_{ij}}{\partial \hat{\underline{d}}^r_{ij}}; \tag{35}$$

$$\frac{\partial \bar{z}_k}{\partial \hat{\bar{c}}^l_{ij}} = \frac{\partial \bar{z}_k}{\partial \bar{\mu}_{ij}} \frac{\partial \bar{\mu}_{ij}}{\partial \hat{\bar{c}}^l_{ij}} = \frac{\bar{z}_k}{\bar{\mu}_{ij}} \frac{\partial \bar{\mu}_{ij}}{\partial \hat{\bar{c}}^l_{ij}}; \quad \frac{\partial \underline{z}_k}{\partial \hat{\bar{c}}^l_{ij}} = 0; \tag{36}$$

$$\frac{\partial \bar{z}_k}{\partial \hat{\bar{c}}^r_{ij}} = \frac{\partial \bar{z}_k}{\partial \bar{\mu}_{ij}} \frac{\partial \bar{\mu}_{ij}}{\partial \hat{\bar{c}}^r_{ij}} = \frac{\bar{z}_k}{\bar{\mu}_{ij}} \frac{\partial \bar{\mu}_{ij}}{\partial \hat{\bar{c}}^r_{ij}}; \quad \frac{\partial \underline{z}_k}{\partial \hat{\bar{c}}^r_{ij}} = 0; \tag{37}$$

$$\frac{\partial \bar{z}_k}{\partial \hat{\bar{d}}^l_{ij}} = \frac{\partial \bar{z}_k}{\partial \bar{\mu}_{ij}} \frac{\partial \bar{\mu}_{ij}}{\partial \hat{\bar{d}}^l_{ij}} = \frac{\bar{z}_k}{\bar{\mu}_{ij}} \frac{\partial \bar{\mu}_{ij}}{\partial \hat{\bar{d}}^l_{ij}}; \quad \frac{\partial \underline{z}_k}{\partial \hat{\bar{d}}^l_{ij}} = 0; \tag{38}$$

$$\frac{\partial \bar{z}_k}{\partial \hat{\bar{d}}^r_{ij}} = \frac{\partial \bar{z}_k}{\partial \bar{\mu}_{ij}} \frac{\partial \bar{\mu}_{ij}}{\partial \hat{\bar{d}}^r_{ij}} = \frac{\bar{z}_k}{\bar{\mu}_{ij}} \frac{\partial \bar{\mu}_{ij}}{\partial \hat{\bar{d}}^r_{ij}}; \quad \frac{\partial \underline{z}_k}{\partial \hat{\bar{d}}^r_{ij}} = 0; \tag{39}$$

In (32)–(39), the derivative terms $\frac{\partial \underline{\mu}_{ij}}{\partial \hat{\underline{c}}^l_{ij}}, \frac{\partial \underline{\mu}_{ij}}{\partial \hat{\underline{c}}^r_{ij}}, \frac{\partial \underline{\mu}_{ij}}{\partial \hat{\underline{d}}^l_{ij}}, \frac{\partial \underline{\mu}_{ij}}{\partial \hat{\underline{d}}^r_{ij}}$ and $\frac{\partial \bar{\mu}_{ij}}{\partial \hat{\bar{c}}^l_{ij}}, \frac{\partial \bar{\mu}_{ij}}{\partial \hat{\bar{c}}^r_{ij}}, \frac{\partial \bar{\mu}_{ij}}{\partial \hat{\bar{d}}^l_{ij}}, \frac{\partial \bar{\mu}_{ij}}{\partial \hat{\bar{d}}^r_{ij}}$ are calculated by considering the input regions, as shown in Tables 1 and 2.

Applying the adaptation laws given in (22)–(31), the estimation controller $\hat{\boldsymbol{u}}_{SEIT2F-AMF}$ can online adjust its parameters.

Table 1 The derivative terms for updating the lower membership function parameters

Region I $x_i \le \hat{\underline{c}}^l_{ij}$	Region II $\hat{\underline{c}}^l_{ij} \le x_i \le \hat{\underline{c}}^r_{ij}$	Region III $\hat{\underline{c}}^r_{ij} \le x_i$
$\frac{\partial \underline{\mu}_{ij}}{\partial \hat{\underline{c}}^l_{ij}} = \underline{\mu}_{ij} \frac{\left(x_i - \hat{\underline{c}}^l_{ij}\right)}{\left(\hat{\underline{d}}^l_{ij}\right)^2}$; $\frac{\partial \underline{\mu}_{ij}}{\partial \hat{\underline{d}}^r_{ij}} = 0$;	$\frac{\partial \underline{\mu}_{ij}}{\partial \hat{\underline{c}}^l_{ij}} = 0$; $\frac{\partial \underline{\mu}_{ij}}{\partial \hat{\underline{d}}^r_{ij}} = 0$;	$\frac{\partial \underline{\mu}_{ij}}{\partial \hat{\underline{c}}^l_{ij}} = 0$; $\frac{\partial \underline{\mu}_{ij}}{\partial \hat{\underline{d}}^r_{ij}} = \underline{\mu}_{ij} \frac{\left(x_i - \hat{\underline{c}}^r_{ij}\right)^2}{\left(\hat{\underline{d}}^r_{ij}\right)^3}$;
$\frac{\partial \underline{\mu}_{ij}}{\partial \hat{\underline{c}}^r_{ij}} = 0$; $\frac{\partial \underline{\mu}_{ij}}{\partial \hat{\underline{d}}^l_{ij}} = \underline{\mu}_{ij} \frac{\left(x_i - \hat{\underline{c}}^l_{ij}\right)^2}{\left(\hat{\underline{d}}^l_{ij}\right)^3}$;	$\frac{\partial \underline{\mu}_{ij}}{\partial \hat{\underline{c}}^r_{ij}} = 0$; $\frac{\partial \underline{\mu}_{ij}}{\partial \hat{\underline{d}}^l_{ij}} = 0$;	$\frac{\partial \underline{\mu}_{ij}}{\partial \hat{\underline{c}}^r_{ij}} = \underline{\mu}_{ij} \frac{\left(x_i - \hat{\underline{c}}^r_{ij}\right)}{\left(\hat{\underline{d}}^r_{ij}\right)^2}$; $\frac{\partial \underline{\mu}_{ij}}{\partial \hat{\underline{d}}^l_{ij}} = 0$;

Table 2 The derivative terms for updating the upper membership function parameters

Region I $x_i \le \hat{\bar{c}}^l_{ij}$	Region II $\hat{\bar{c}}^l_{ij} \le x_i \le \hat{\bar{c}}^r_{ij}$	Region III $\hat{\bar{c}}^r_{ij} \le x_i$
$\frac{\partial \overline{\mu}_{ij}}{\partial \hat{\bar{c}}^l_{ij}} = \overline{\mu}_{ij} \frac{\left(x_i - \hat{\bar{c}}^l_{ij}\right)}{\left(\hat{\bar{d}}^l_{ij}\right)^2}; \frac{\partial \overline{\mu}_{ij}}{\partial \hat{\bar{d}}^r_{ij}} = 0;$ $\frac{\overline{\mu}_{ij}}{\partial \hat{\bar{c}}^r_{ij}} = 0; \frac{\partial \overline{\mu}_{ij}}{\partial \hat{\bar{d}}^l_{ij}} = \overline{\mu}_{ij} \frac{\left(x_i - \hat{\bar{c}}^l_{ij}\right)^2}{\left(\hat{\bar{d}}^l_{ij}\right)^3};$	$\frac{\partial \overline{\mu}_{ij}}{\partial \hat{\bar{c}}^l_{ij}} = 0; \frac{\partial \overline{\mu}_{ij}}{\partial \hat{\bar{d}}^r_{ij}} = 0;$ $\frac{\overline{\mu}_{ij}}{\partial \hat{\bar{c}}^r_{ij}} = 0; \frac{\partial \overline{\mu}_{ij}}{\partial \hat{\bar{d}}^l_{ij}} = 0;$	$\frac{\partial \overline{\mu}_{ij}}{\partial \hat{\bar{c}}^l_{ij}} = 0; \frac{\partial \overline{\mu}_{ij}}{\partial \hat{\bar{d}}^r_{ij}} = \overline{\mu}_{ij} \frac{\left(x_i - \hat{\bar{c}}^r_{ij}\right)^2}{\left(\hat{\bar{d}}^r_{ij}\right)^3};$ $\frac{\overline{\mu}_{ij}}{\partial \hat{\bar{c}}^r_{ij}} = \overline{\mu}_{ij} \frac{\left(x_i - \hat{\bar{c}}^r_{ij}\right)}{\left(\hat{\bar{d}}^r_{ij}\right)^2}; \frac{\partial \overline{\mu}_{ij}}{\partial \hat{\bar{d}}^l_{ij}} = 0;$

3.2 *Structure Learning Algorithm*

For the design of fuzzy neural network, determination of the network size (the number of rules) is much effect to the system performance. More rules lead to huge computation time and cause the controller to respond slowly. Contrarily, few rules may not cover all cases and cause the controller cannot perform well. In this section, the self-evolving algorithm is applied to autonomously construct the network structure of the proposed SEIT2F-AMF controller.

A new rule will be generated if the following condition is satisfied:

$$\mu^i_{\max} < T_g \tag{40}$$

where $\mu^i_{\max} = \max\left[\mu_{i1}, \mu_{i2}, \ldots, \mu_{in_j}\right]$ is the maximum membership grade for ith input; T_g is the generating threshold; $\mu_{ij} = \frac{\underline{\mu}_{ij} + \overline{\mu}_{ij}}{2}$ is the average of the upper and lower membership function.

If (40) is satisfied, the initial variance and mean of the new AMF will be given as:

$$\left[\bar{c}^l_{ij}, \underline{c}^l_{ij}, \underline{c}^r_{ij}, \bar{c}^r_{ij}\right] = [(x_i - 2\lambda), (x_i - \lambda), (x_i + \lambda), (x_i + 2\lambda)] \tag{41}$$

$$\left[\underline{d}^l_{ij}, \bar{d}^l_{ij}, \underline{d}^r_{ij}, \bar{d}^r_{ij}\right] = \left[(d_{init} - 2\gamma), (d_{init} - \gamma), (d_{init} + \gamma), (d_{init} + 2\gamma)\right] \tag{42}$$

where λ and γ are the parameters for adjusting the mean and variance of AMF; d_{init} is the initial variance.

An unused rule will be deleted if the following condition is satisfied:

$$\mu^i_{\min} < T_d \tag{43}$$

where $\mu^i_{\min} = min\left[\mu_{i1}, \mu_{i2}, \ldots, \mu_{in_j}\right]$ is the minimum membership grade for ith input; T_d is the deleting threshold.

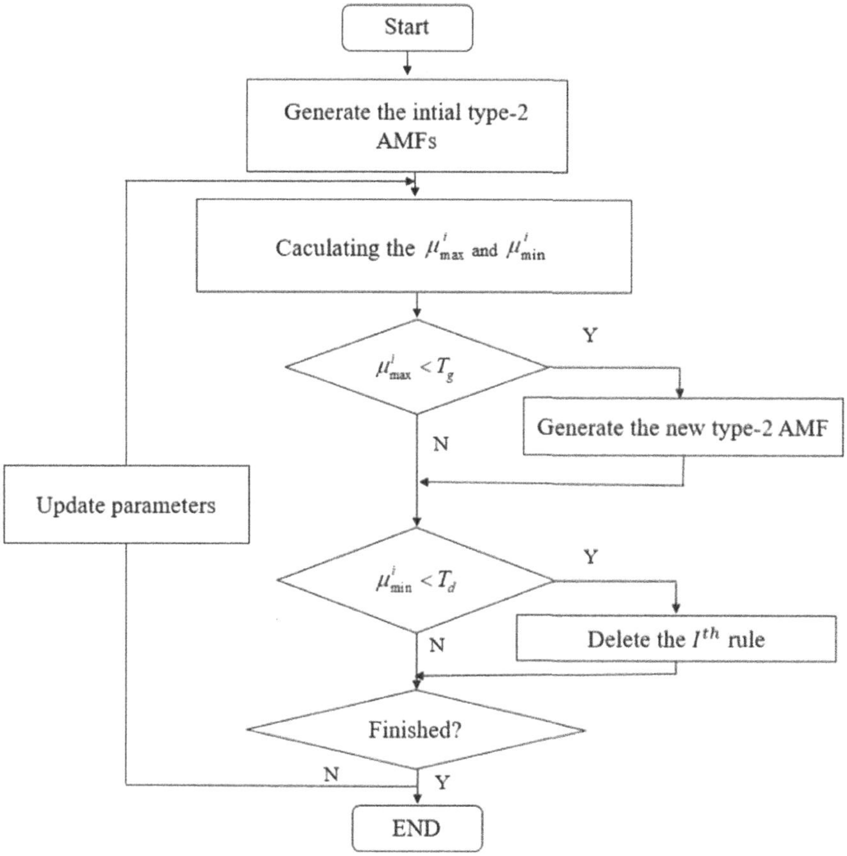

Fig. 4 The flowchart of structure learning algorithm

If (43) is satisfied, the minimum membership grade and its corresponding AMF will be deleted.

Applying the proposed self-evolving algorithm, the SEIT2F-AMF controller can achieve a suitable structure. The flowchart of structure learning is given in Fig. 4.

4 Simulation Results

A four-dimensional Lorenz–Stenflo chaotic system in [62] can be described as follows:

$$\dot{x}_1(t) = h(x_2(t) - x_1(t)) + px_4$$
$$\dot{x}_2(t) = rx_1(t) - x_1(t)x_3(t) - x_2(t)$$

$$\begin{aligned}\dot{x}_3(t) &= x_1(t)x_2(t) - vx_3(t)\\ \dot{x}_4(t) &= -x_1(t) - hx_4\end{aligned} \tag{44}$$

$$\begin{aligned}\dot{y}_1(t) &= h(y_2(t) - y_1(t)) + py_4 + \upsilon_1(t) + \Delta f(y_1) + u_1(t)\\ \dot{y}_2(t) &= ry_1(t) - y_1(t)y_3(t) - y_2(t) + \upsilon_2(t) + \Delta f(y_2) + u_2(t)\\ \dot{y}_3(t) &= y_1(t)y_2(t) - vy_3(t) + \upsilon_3(t) + \Delta f(y_3) + u_3(t)\\ \dot{y}_4(t) &= -y_1(t) - hy_4 + \upsilon_4(t) + \Delta f(y_4) + u_4(t)\end{aligned} \tag{45}$$

where $\boldsymbol{x}(t) = [x_1(t), x_2(t), x_3(t), x_4(t)]$ and $\boldsymbol{y}(t) = [y_1(t), y_2(t), y_3(t), y_4(t)]$ are the master and slave system, respectively; $\boldsymbol{\upsilon}(t) = [\boldsymbol{\upsilon}_1(t), \boldsymbol{\upsilon}_2(t), \boldsymbol{\upsilon}_3(t), \boldsymbol{\upsilon}_4(t)]$ and $\Delta\boldsymbol{f}(t) = [\Delta f(y_1), \Delta f(y_2), \Delta f(y_3), \Delta f(y_4)]$ are the external disturbances and the uncertainties vector, respectively; $\boldsymbol{u}(t) = [u_1(t), u_2(t), u_3(t), u_4(t)]$ is the control vector for chaotic system synchronization; h, p, r, v denote the parameters for defining the chaotic attractor.

The error dynamics between (47) and (48) are given as:

$$\begin{aligned}e_1(t) &= y_1(t) - x_1(t)\\ e_2(t) &= y_2(t) - x_2(t)\\ e_3(t) &= y_3(t) - x_3(t)\\ e_4(t) &= y_4(t) - x_4(t)\end{aligned} \tag{46}$$

The initial position states of the master system and slave system are chosen as $x_1 = 0.028$, $x_2 = 0.02$, $x_3 = 0.03$, $x_4 = 0.048$ and $y_1 = 0.01$, $y_2 = 0.037$, $y_3 = 0.029$, $y_4 = 0.008$. The external disturbances and system uncertainties are chosen as $\upsilon_1(t) = 0.2\cos\pi t$, $\upsilon_2(t) = 0.5\cos\pi t$, $\upsilon_3(t) = 0.3\cos\pi t$, $\upsilon_4(t) = 0.4\cos\pi t$ and $\Delta f(y_1) = rd(.)0.2y_1$, $\Delta f(y_2) = rd(.)0.2y_2$, $\Delta f(y_3) = rd(.)0.2y_3$, $\Delta f(y_4) = rd(.)0.2y_4$, where $rd(.)$ denotes the random number in range [0, 1]. The learning rates are chosen as $\hat{\eta}_q = 0.06$, $\hat{\eta}_c = 0.01$, $\hat{\eta}_d = 0.0001$. The thresholds for structure learning algorithms are chosen as $T_g = 0.01$ and $T_d = 0.001$. The initial parameters for generating new AMF are $d_{init} = 0.3$, $\lambda = 0.1$ and $\gamma = 0.05$.

The parameters for chaotic system are set as $h = 1/2$, $v = 8/3$, $p = 20$, $r = 30$. The phase portraits for four-dimensional Lorenz–Stenflo chaotic system are shown in Fig. 5. The state trajectories of the proposed SEIT2F-AMF chaotic system synchronization are given in Fig. 6. The control efforts $u_1(t)$, $u_2(t)$, $u_3(t)$, $u_4(t)$ and tracking errors $e_1(t)$, $e_2(t)$, $e_3(t)$, $e_4(t)$ are respectively shown in Figs. 7 and 8. The online adjustment number of rules are shown in Fig. 9. The simulation results show that the proposed SEIT2F-AMF controller can synchronize the four-dimensional Lorenz–Stenflo chaotic system well even with the external disturbances and system uncertainties. The root mean square error (RMSE) in this case is 0.2102.

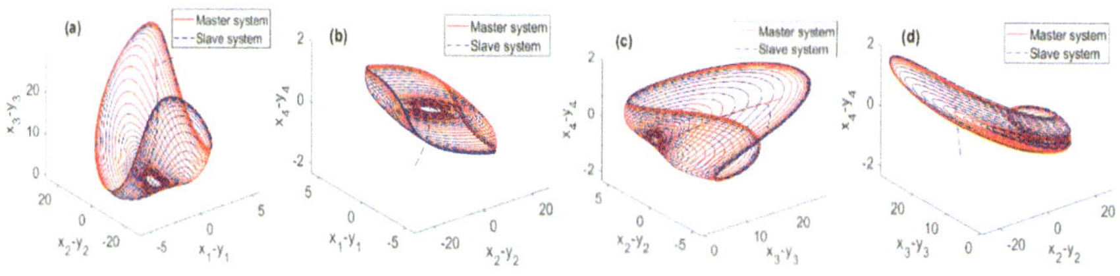

Fig. 5 The phase portraits for four-dimensional Lorenz–Stenflo chaotic system: **a** $x_1 - x_2 - x_3$, **b** $x_2 - x_1 - x_4$, **c** $x_3 - x_2 - x_4$, **d** $x_2 - x_3 - x_4$

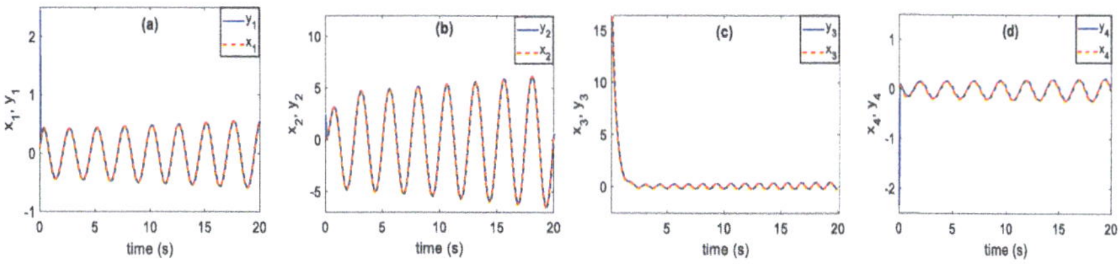

Fig. 6 The time response state trajectories of the proposed SEIT2F-AMF chaotic synchronization: **a** x_1, y_1, **b** x_2, y_2, **c** x_3, y_3, and **d** x_4, y_4

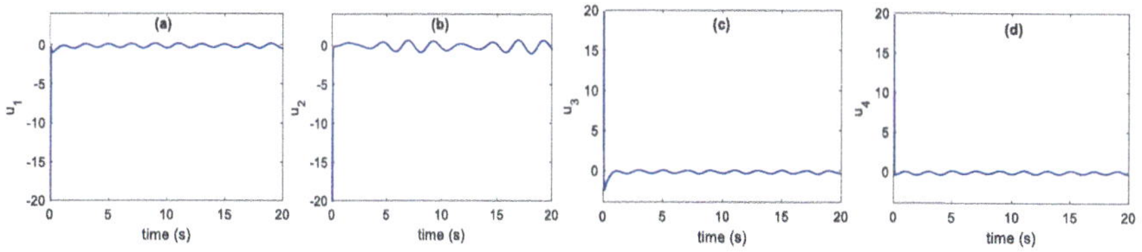

Fig. 7 The control efforts: **a** u_1, **b** u_2, **c** u_3, and **d** u_4

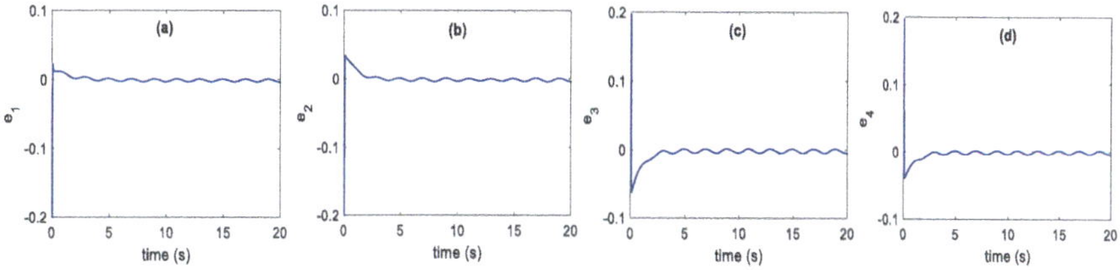

Fig. 8 The tracking errors: **a** e_1, **b** e_2, **c** e_3, and **d** e_4

5 Conclusion

In this study, a SEIT2F-AMF controller is designed for the synchronization of the four-dimensional Lorenz–Stenflo chaotic system. The function-link network and the asymmetric membership function are applied to enhance the learning capability and

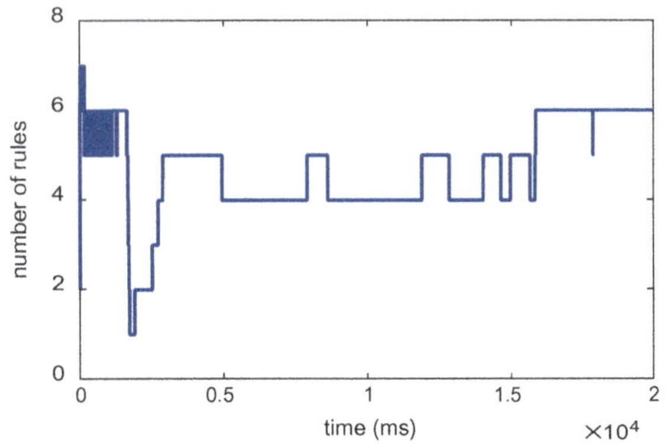

Fig. 9 The online adjustments number of rules

flexibility of the proposed network. The adaptation laws for online updating network parameters are designed using gradient descent method. In addition, the network structure can autonomously be constructed by using the self-evolving algorithm. The simulation results show that the proposed method can synchronize the four-dimensional Lorenz–Stenflo chaotic system well even with the external disturbances and system uncertainties. For future researches, optimization of learning rates for adaptation laws and optimization of threshold for structure learning algorithm can be studied.

References

1. Wang, Y., Zhang, X., Yang, L., Huang, H.: Adaptive synchronization of time delay chaotic systems with uncertain and unknown parameters via aperiodically intermittent control. Int. J. Control Autom. Syst. **18**(3), 696–707 (2020)
2. Ahmad, I., Shafiq, M.: Oscillation free robust adaptive synchronization of chaotic systems with parametric uncertainties. Trans. Inst. Measur. Control (2020). 0142331220903668
3. Boulkroune, A., Bouzeriba, A., Bouden, T.: Fuzzy generalized projective synchronization of incommensurate fractional-order chaotic systems. Neurocomputing **173**, 606–614 (2016)
4. Mohammadzadeh, A., Ghaemi, S.: Optimal synchronization of fractional-order chaotic systems subject to unknown fractional order, input nonlinearities and uncertain dynamic using type-2 fuzzy CMAC. Nonlinear Dyn. **88**(4), 2993–3002 (2017)
5. Mufti, M.R., Afzal, H., Rehman, F.U., Butt, Q.R., Qureshi, M.I.J.I.A.: Synchronization and antisynchronization between two non-identical Chua oscillators via sliding mode control **6**, 45270–45280 (2018)
6. Xu, L., Ma, H., Xiao, S.: Exponential synchronization of chaotic Lur'e systems using an adaptive event-triggered mechanism. IEEE Access **6**, 61295–61304 (2018)
7. Li, S.-Y., Hernández, M.A.B.: Robust synchronization of chaotic systems with novel fuzzy rule-based controllers. Inf. Sci. **481**, 604–615 (2019)
8. Le, T.-L., Lin, C.-M., Huynh, T.-T.: Interval type-2 petri CMAC design for 4D chaotic system. In: 2019 International Conference on System Science and Engineering (ICSSE). Dong Hoi, Quang Binh, Vietnam: IEEE. pp. 420–4

9. Li, X., Ma, D., Xie, X., Sun, Q.: Fault-tolerant synchronization of chaotic systems with fuzzy sampled data controller based on adaptive event-triggered scheme. Int. J. Fuzzy Syst. 1–13 (2020)
10. Le, T.-L., Huynh, T.-T., Hong, S.-K.: Self-organizing interval type-2 fuzzy asymmetric CMAC design to synchronize chaotic satellite systems using a modified grey wolf optimizer. IEEE Access **8**, 53697–53709 (2020)
11. Modiri, A., Mobayen. S.: Adaptive terminal sliding mode control scheme for synchronization of fractional-order uncertain chaotic systems. ISA Trans. (2020)
12. Le, T.-L., Huynh, T.-T., Hong, S.K.: A modified grey wolf optimizer for optimum parameters of multilayer type-2 asymmetric fuzzy controller. IEEE Access **8**, 121611–121629 (2020)
13. Wang, R., Zhang, Y., Chen, Y., Chen, X., Xi, L.: Fuzzy neural network-based chaos synchronization for a class of fractional-order chaotic systems: an adaptive sliding mode control approach. Nonlinear Dyn. 1–13 (2020)
14. Le, T.-L., Huynh, T.-T., Nguyen, V.-Q., Lin, C.-M., Hong, S.-K.: Chaotic synchronization using a self-evolving recurrent interval type-2 petri cerebellar model articulation controller. Mathematics **8**(2), 219 (2020)
15. Wang, Y., Tong, D., Chen, Q., Zhou, W.: Exponential synchronization of chaotic systems with stochastic perturbations via quantized feedback control. Circuits Syst. Sign. Process. **39**(1), 474–491 (2020)
16. Gil, P., Oliveira, T., Palma, L.B.: Online non-affine nonlinear system identification based on state-space neuro-fuzzy models. Soft. Comput. **23**(16), 7425–7438 (2019)
17. Huang, S., Zhao, G., Chen, M.: Uniform design-based interval type-2 neuro-fuzzy system and its performance verification. Int. J. Fuzzy Syst. **20**(6), 1821–1838 (2018)
18. Zadeh, L.A.: Fuzzy sets. Inf. Control **8**(3), 338–353 (1965)
19. Zhou, Y., Miao, F., Luo, Q.: Symbiotic organisms search algorithm for optimal evolutionary controller tuning of fractional fuzzy controllers. Appl. Soft Comput. **77**, 497–508 (2019)
20. Farajdadian, S., Hosseini, S.H.: Design of an optimal fuzzy controller to obtain maximum power in solar power generation system. Sol. Energy **182**, 161–178 (2019)
21. Han, H., Wu, X., Liu, Z., Qiao, J.: Data-knowledge-based fuzzy neural network for nonlinear system identification. IEEE Trans. Fuzzy Syst. (2019). https://doi.org/10.1109/TFUZZ.2019.2931870
22. Škrjanc, I., Iglesias, J.A., Sanchis, A., Leite, D., Lughofer, E., Gomide, F.: Evolving fuzzy and neuro-fuzzy approaches in clustering, regression, identification, and classification: a survey. Inf. Sci. **490**, 344–368 (2019)
23. Xu, S., Liu, K., Li, X.: A fuzzy process neural network model and its application in process signal classification. Neurocomputing **335**, 1–8 (2019)
24. Chang, Y., Wang, Y., Alsaadi, F.E., Zong, G.: Adaptive fuzzy output-feedback tracking control for switched stochastic pure-feedback nonlinear systems. Int. J. Adapt. Control Sign. Process. **33**(10), 1567–1582 (2019)
25. Marín, L.G., Cruz, N., Sáez, D., Sumner, M., Núñez, A.: Prediction interval methodology based on fuzzy numbers and its extension to fuzzy systems and neural networks. Expert Syst. Appl. **119**, 128–141 (2019)
26. Tsakiridis, N.L., Theocharis, J.B., Panagos, P., Zalidis, G.C.: An evolutionary fuzzy rule-based system applied to the prediction of soil organic carbon from soil spectral libraries. Appl. Soft Comput. **81**:105504 (2019)
27. Aliev, R.A., Guirimov, B.G.: Type-2 Fuzzy Neural Networks and Their Applications. Springer (2014)
28. Lin, C.-M., La, V.-H., Le, T.-L.: DC–DC converters design using a type-2 wavelet fuzzy cerebellar model articulation controller. Neural Comput. Appl. **32**(7), 2217–2229 (2020)
29. Le, T.-L.: Fuzzy c-means clustering interval type-2 cerebellar model articulation neural network for medical data classification. IEEE Access **7**, 20967–20973 (2019)
30. Zadeh, L.A.: The concept of a linguistic variable and its application to approximate reasoning—I. Inf. Sci. **8**(3), 199–249 (1975)

31. Liang, Q., Mendel, J.M.: Interval type-2 fuzzy logic systems: theory and design. IEEE Trans. Fuzzy Syst. **8**(5), 535–550 (2000)
32. Eyoh, I., John, R., De Maere, G.: Interval type-2 intuitionistic fuzzy logic for regression problems. IEEE Trans. Fuzzy Syst. (2017)
33. Huang, J., Ri, M., Wu, D., Ri, S.: Interval type-2 fuzzy logic modeling and control of a mobile two-wheeled inverted pendulum. IEEE Trans. Fuzzy Syst. **26**(4), 2030–2038 (2017)
34. Le, T.-L., Lin, C.-M., Huynh, T.-T.: Self-evolving type-2 fuzzy brain emotional learning control design for chaotic systems using PSO. Appl. Soft Comput. **73**, 418–433 (2018)
35. Amador-Angulo, L., Castillo, O.: A new fuzzy bee colony optimization with dynamic adaptation of parameters using interval type-2 fuzzy logic for tuning fuzzy controllers. Soft. Comput. **22**(2), 571–594 (2018)
36. Le, T.-L.: Self-organizing recurrent interval type-2 Petri fuzzy design for time-varying delay systems. IEEE Access **7**, 10505–10514 (2018)
37. Zhao, T., Liu, J., Dian, S.: Finite-time control for interval type-2 fuzzy time-delay systems with norm-bounded uncertainties and limited communication capacity. Inf. Sci. **483**, 153–173 (2019)
38. Le, T.-L., Huynh, T.-T., Lin, L.-Y., Lin, C.-M., Chao, F.: A k-means interval type-2 fuzzy neural network for medical diagnosis. Int. J. Fuzzy Syst. **21**(7), 2258–2269 (2019)
39. Méndez, G.M., Dorantes, P.N.M., Santoyo, A.M.: Interval type-2 fuzzy logic systems optimized by central composite design to create a simplified fuzzy rule base in image processing for quality control application. Int. J. Adv. Manuf. Technol. **102**(9–12), 3757–3766 (2019)
40. Le, T.-L.: Intelligent fuzzy controller design for antilock braking systems. J. Intell. Fuzzy Syst. **36**(4), 3303–3315 (2019)
41. Dey, A., Pal, A., Long, H.V.: Fuzzy minimum spanning tree with interval type 2 fuzzy arc length: formulation and a new genetic algorithm. Soft. Comput. **24**(6), 3963–3974 (2020)
42. Qin, J., Xi, Y., Pedrycz, W.: Failure mode and effects analysis (FMEA) for risk assessment based on interval type-2 fuzzy evidential reasoning method. Appl. Soft Comput. **89**:106134 (2020)
43. Ontiveros, E., Melin, P., Castillo, O.: Comparative study of interval Type-2 and general Type-2 fuzzy systems in medical diagnosis. Inf. Sci. **525**, 37–53 (2020)
44. Chen, S.-Y., Liu, T.-S.: Intelligent tracking control of a PMLSM using self-evolving probabilistic fuzzy neural network. IET Electr. Power Appl. **11**(6), 1043–1054 (2017)
45. Lin, C.-M., Le, T.-L., Huynh, T.-T.: Self-evolving function-link interval type-2 fuzzy neural network for nonlinear system identification and control. Neurocomputing **275**, 2239–2250 (2018)
46. Le, T.-L., Huynh, T.-T., Lin, C.-M.: Self-evolving interval type-2 wavelet cerebellar model articulation control design for uncertain nonlinear systems using PSO. Int. J. Fuzzy Syst. **21**(8), 2524–2541 (2019)
47. Ge, D., Zeng, X.-J.: Learning data streams online—an evolving fuzzy system approach with self-learning/adaptive thresholds. Inf. Sci. **507**, 172–184 (2020)
48. Afrakhte, H., Bayat, P.: A self-evolving type-2 fuzzy energy management strategy for multi-microgrid systems. Comput. Electr. Eng. **85**:106702 (2020)
49. Pao, Y.-H.: Functional link nets: removing hidden layers. AI Expert. **4**(4), 60–68 (1989)
50. Lotfi, E., Rezaee, A.A.: A competitive functional link artificial neural network as a universal approximator. Soft. Comput. **22**, 4613–4625 (2018)
51. Zhang, L., Suganthan, P.N.: A comprehensive evaluation of random vector functional link networks. Inf. Sci. **367**, 1094–1105 (2016)
52. Scardapane, S., Wang, D., Uncini, A.: Bayesian random vector functional-link networks for robust data modeling. IEEE Trans. Cybern. **48**(7), 2049–2059 (2017)
53. Li, L., Qu, X., Zhang, J., Li, H., Ran, B.: Travel time prediction for highway network based on the ensemble empirical mode decomposition and random vector functional link network. Appl. Soft Comput. **73**, 921–932 (2018)
54. Katuwal, R., Suganthan, P.N., Zhang, L.: An ensemble of decision trees with random vector functional link networks for multi-class classification. Appl. Soft Comput. **70**, 1146–1153 (2018)

55. Zhou, Q., Chao, F., Lin, C.-M.: A functional-link-based fuzzy brain emotional learning network for breast tumor classification and chaotic system synchronization. Int. J. Fuzzy Syst. **20**(2), 349–365 (2018)
56. Huynh, T.-T., Lin, C.-M., Pham, T.-T.T., Cho, H.-Y., Le, T.-L.: A modified function-link fuzzy cerebellar model articulation controller using a PI-type learning algorithm for nonlinear system synchronization and control. Chaos, Solitons Fractals **118**:65–82 (2019)
57. Zhang, P.-B., Yang, Z.-X.: A new learning paradigm for random vector functional-link network: RVFL+. Neural Netw. **122**, 94–105 (2020)
58. Zhao, T., Li, P., Cao, J.: Self-organising interval type-2 fuzzy neural network with asymmetric membership functions and its application. Soft. Comput. **23**(16), 7215–7228 (2019)
59. Tan, K.-H., Lin, F.-J., Chen, J.-H.: DC-link voltage regulation using RPFNN-AMF for three-phase active power filter. IEEE Access **6**, 37454–37463 (2018)
60. Lin, F.-J., Tan, K.-H., Lai, Y.-K., Luo, W.-C.: Intelligent PV power system with unbalanced current compensation using CFNN-AMF. IEEE Trans. Power Electron. **34**(9), 8588–8598 (2018)
61. Mendel, J.M.: Uncertain rule-based fuzzy logic systems: introduction and new directions. Prentice Hall PTR Upper Saddle River (2001)
62. Vincent, U.: Synchronization of identical and non-identical 4-D chaotic systems using active control. Chaos, Solitons Fractals **37**(4), 1065–1075 (2008)

Categorizing Relations via Semi-supervised Learning Using a Hybrid Tolerance Rough Sets and Genetic Algorithm Approach

Shubham Agrawal, Rashad Ahmed, M. Anand Kumar, and Sheela Ramanna

Abstract In the last few decades, we have seen a tremendous increase in the amount of data available on the web. There have been significant advances in constructing knowledge bases consisting of relations from the text data. These relations are words in the text often represented as pairs (Noun, Context), for example (Disease, Symptom), which can be classified into some predefined category to give us some useful information. Categorization of relations using tolerance-rough set based semi-supervised learning algorithm (TPL) have been successfully demonstrated in several works. However, an unexplored problem is the automatic selection of hyper parameters of the TPL algorithm. This paper proposes a genetic algorithm-based approach (TPL-GA) for optimizing the hyper-parameters that are fundamental to the TPL algorithm. The proposed approach was tested on two standard datasets drawn from different domains representing two different languages: English and Hindi text.

Keywords Co-occurrence matrix · Genetic algorithm · Semi-supervised learning · Tolerance rough sets · Tolerant pattern learner (TPL)

1 Introduction

The amount of text generated is enormous, and using it to gather some useful information is vital. The task here is to use the existing text data, which is mostly unstructured and label it to extract some information and gain knowledge. Extracting information and gaining knowledge can be done by categorizing the text into some categories. The text data usually contains entities such as the name of a person, country or various diseases. Along with these entities, the text also contains information (context) which

Shubham Agrawal · Rashad Ahmed · M. Anand Kumar (✉)
Department of Information Technology, National Institute of Technology Karnataka, Surathkal, India
e-mail: m_anandkumar@nitk.edu.in

Sheela Ramanna
Department of Applied Computer Science, University of Winnipeg, Winnipeg, Canada
e-mail: s.ramanna@uwinnipeg.ca

D. Gupta et al. (eds.), *Soft Computing for Data Analytics, Classification Model, and Control*, Studies in Fuzziness and Soft Computing 413,
https://doi.org/10.1007/978-3-030-92026-5_6

define this entity, such as, a person can be a famous politician or a sports player, or a disease could contain some specific symptoms related to it. The entities can be a unary relation (noun) or a binary relation (relation). The contextual information or patterns aid in the categorization of relations into predefined categories.

In this paper, we use a semi-supervised learning algorithm based soft computing technique (tolerance rough sets) as explained in [1, 2], and introduce a hybrid method based on Genetic Algorithm [3]. Tolerance form of rough sets have been used extensively for document and text processing [4]. While classical rough sets rely on an equivalence relation [5], tolerance rough sets are based on a tolerance relation which is symmetric and reflexive but not transitive [6]. This property is used in the TPL algorithm to represent overlapping classes which is then used to extract knowledge from text-based entities. The TPL algorithm takes text-based data as input along with some predefined categories. This algorithm works iteratively, which has a significant advantage as it uses the knowledge generated from the previous iteration in the next iteration. TPL discovers binary relations within the text such as between a noun and a verb like (kangaroo, animal) or disease and its symptoms like (asthma, cough). A co-occurrence matrix is formed between these pairs, which gives co-occurrence of entities and it's corresponding context in the data. Each cell in the matrix gives the exact number of such occurrences, which then helps find the association between similar entities, thus categorizing them.

TPL uses the tolerance approximation space to create an enriched representation for the noun phrases that are promoted. Each promoted noun phrase is associated with three different descriptive sets. These sets of the trusted noun instance are used to calculate the micro score for other candidate nouns. Here, the micro scores are calculated using a particular contribution from each of the three sets. The TPL algorithm learns by separating the tolerance approximation space for each entity in the text-data that we want to classify into different zones, where each zone contributes to a different degree of similarity for each entity. The approximation space is divided into four zones. The lower and upper approximators are used to calculate the micro score. Here, some hyper parameters known as weights are used that reflect the degree of similarity that is contributed by the entity belonging to that specific zone. In the current implementation of TPL, these hyper parameters are determined by manually. This could pose a challenge when learning from large datasets.

It is this limitation that we address in the proposed TPL-GA method by using a Genetic algorithm [3]. Genetic algorithm is a widely used algorithm for optimization purposes. It is a search heuristic inspired by the natural evolution phenomenon described by Charles Darwin's theory. The GA algorithm randomly initializes a population of random values between a particular range. Operations such as fitness score calculation, offspring crossover, mutation are used to find the most suitable hyper parameter values for the micro score calculation. There are certain challenges while trying to categorize the text-data. Sometimes, the size of a particular text file might be small, and the number of training examples might be less, for which we are using a semi-supervised learning-based algorithm which is optimized using the genetic algorithm. It requires some seed examples per category to be fed to the algorithm, which then it uses to classify other instances iteratively as mentioned earlier.

Another problem is that the algorithm uses a co-occurrence matrix, which plays a crucial factor in algorithms accuracy. Because of this, even a slightest of variation in the matrix can give varying results. We have also automated the matrix construction based on the type of text used. Finally, the main contribution of this work is the introduction of a Hybrid TPL-GA algorithm that leverages the optimization capabilities of the genetic algorithm to optimize the TPL hyper parameters. The TPL-GA algorithm has shown significant improvement in the categorization accuracy not only with English text but also in categorizing texts in a second domain (health) and natural language (Hindi).

2 Literature Survey

To solve the task of extracting knowledge from text-based data, there are many successful Information Extraction systems that use machine learning such as YAGO3 [7–9], KnowItAll [10, 11], TextRunner [10, 12]. Never-ending learner (NELL) [13], includes Coupled Pattern Learner (CPL) [2] and Coupled Bayesian Sets (CBS) [14] algorithms for learning relational patterns. Existing techniques trying to solve this task can mainly be classified into two categories: statistical information-based approaches and machine learning-based methods. The first one usually relies on manually designed extraction rules, and it uses noun phrases of high frequency as the candidate product features. However, one drawback is that it can produce many non-product features and could miss many low-frequency terms. However, the ML techniques try to treat this problem as a probabilistic learning process.

TPL [1] has its roots in rough set theory which consists of an approximation space that is characterized by an equivalence relation. It learns using two approximation operators: lower and upper approximation. The main assumption of rough set theory is that with every object/entity of the universe, we can gather or associate some information (which here is referred to as data, knowledge). All the objects characterized by the same information are similar given the available information about them. This relation is the mathematical basis of rough set theory. Any such set of all indiscernible objects is called an elementary set and forms a basic granule of knowledge about the universe. More recently, rough set [1] and fuzzy rough set-based [15] algorithms for categorizing relations were introduced as semi-supervised algorithms that require minimal human intervention and are fully automated.

Genetic Algorithm (GA) dates back to 1970, which was introduced by John Holland [3]. GA implements the optimization by simulating the natural evolution of species. It contains processes like selection of the best species and the manipulation of the species to produce better offspring by methods like crossover and mutation [16]. GA is an easy and powerful tool that allows us to identify the best possible value in the search space instead of manually tweaking the equations. A set of chromosomes is called population, where the fitness value is calculated for each chromosome using the objective function value.

3 Problem Statement

The main goal of this paper is to optimize hyper parameters of the TPL algorithm using a classical *genetic algorithm* that mimics the process of natural selection. Specific objectives are as follows:

3.1 Objectives

- To address the task of automating the selection of α, β and γ parameters used in the TPL algorithm.
- To optimize the TPL algorithm in order to make it robust for bigger data sets.
- To demonstrate the efficacy of the TPL-GA algorithm on a small dataset in more than one natural language.

4 Methodology

4.1 Dataset Details

We have used 2 natural language datasets: English (TDIL tourism-2 EILMT corpus—2Tour10k) and Hindi (Hindi Health Dataset—HHD). 2Tour10k consists of 10 K sentences, for which the Parts of Speech (POS) tagging information is first generated. HHD [17, 18] which consists of a list of predefined symptoms and diseases. HHD corpus consists of approximately 6 K sentences and was created using Python 2.7.11 gathered from Indian websites and four gazetteer lists—Person, Disease, Consumable and Symptom. It consists of around 150 diseases and 200 symptoms. The next step is to generate the required co-occurrence matrix.

4.2 Co-occurrence Matrix Generation

Since the TPL algorithm needs an ontology as an input, the first step is to generate such an input from the dataset described in the previous section. After generating the POS tags information for the English dataset, the set of regular expressions are used to extract the required tags as context and members. In our study we assume the members to be nouns. "POS:VERB:? POS:ADV:* POS:VERB:+" is an example of how the context and members are extracted. The matrix entries are filled based on the frequency of occurrence of the members with their corresponding contexts in the dataset. As for the HHD data, the matrix was populated in a similar way, based on the frequency of occurrence. Figure 1 represents the POS tag information for a

```
Tourists|tourist|N|pl can|can|MD|mod see|see|V|nom elegant|elegant|AJ|adj
puppet|puppet|N|nom shows|show|N|pl articulate|articulate|V|nom
fingers|finger|N|pl of|of|P|prep the|the|AR|det charmer|charmer|N|nom
playing|play|V|vb.prog with|with|P|prep the|the|AR|det emotions|emotion|N|pl
of|of|P|prep the|the|AR|det wooden|wooden|AJ|adj toys|toy|N|pl
traditional|traditional|AJ|adj folk|folk|N|nom performances|performance|N|pl
and|and|C|cconj can|can|MD|mod savour|savour|V|nom the|the|AR|det
true|true|AJ|adj taste|taste|N|nom of|of|P|prep Rajasthani|Rajasthani|N|nom
food|food|N|nom .|.|.|.
```

Fig. 1 POS tag info of a sentence from 2Tour10k dataset

sentence, "Tourists can see elegant puppet shows, articulate finger of the charmer playing with the emotions of the wooden toys, traditional folk performances and can savour the true taste of Rajasthani food" from the 2Tour10k dataset. Figure 2 gives a sample co-occurrence matrix obtained for the HHD data. All the rows in the matrix represent the *diseases* while the columns represent the *symptoms*. A total of 36 diseases, with 12 diseases in each category were chosen and more than 60 related symptoms were chosen for our experiments. The diseases were then categorized into three categories, Infectious, Hereditary and Life-style based disease. All the diseases and symptoms used to build the co-occurrence matrix were manually selected and categorized since there is no existing dataset that categorizes the diseases in Hindi text into different categories.

CO-OCCURENCE MATRIX

	ज़ुकाम	कमज़ोरी	जलन	संक्रमण	थकान	खुजली	ठंडा	दस्त	इंफेक्शन	पसीना	चोट	बुखार	वायरल	घाव	फ्लू	कैंसर	गंदगी	तनाव	कमज़ोरी	जलन	एलर्जी
खसरा	1	1	2	4	1	2	1	1	1	1	1	2	1	0	0	0	0	0	1	2	1
फाइलेरिया	0	0	0	1	0	0	0	0	1	0	0	1	0	1	0	0	0	0	0	0	0
निमोनिया	1	0	0	4	4	1	2	0	0	0	1	3	3	1	2	1	0	0	0	0	0
हेपेटाइटिस	0	2	1	2	2	1	0	0	0	0	0	1	2	0	0	3	1	0	2	1	0
डेंगू	1	3	1	2	1	0	2	1	1	1	1	3	2	0	1	0	1	1	3	1	0
मलेरिया	1	4	1	3	2	0	3	1	1	2	1	3	2	0	1	1	2	1	4	1	0
टिटनस	0	0	0	1	0	0	0	0	0	1	1	0	0	1	0	0	0	0	0	0	0
मेनिन्जाइटिस	1	0	0	1	0	0	1	0	1	0	0	1	0	0	0	0	0	0	0	0	0
कुष्ठरोग	0	0	1	0	0	0	0	0	0	0	0	1	0	1	0	0	0	0	0	1	0
टाइफाइड	0	0	0	0	0	0	1	1	1	1	0	1	0	0	0	0	1	0	0	0	0
हैजा	0	0	0	0	0	0	0	1	0	0	0	0	0	0	0	0	1	0	0	0	0
एड्स	1	1	1	2	1	1	1	1	1	1	0	2	1	1	1	0	0	0	1	1	0
दमा	1	1	1	0	0	0	1	0	0	0	0	0	0	0	1	0	1	1	1	1	1
मधुमेह	0	2	2	2	2	1	0	2	2	3	0	1	0	0	1	1	0	4	2	2	2
गठिया	0	2	3	1	0	0	4	2	1	0	4	2	0	0	0	1	0	2	2	3	0
कब्ज	1	3	6	4	4	2	7	5	3	5	2	7	3	0	1	3	4	7	3	6	2
गंजापन	0	0	0	0	0	0	0	0	0	0	0	0	0	0	0	1	0	1	0	0	0
मोटापा	0	3	0	2	2	1	2	0	1	1	3	2	1	0	1	3	1	3	3	0	1
हाइपोथायराइड	0	1	0	0	1	0	1	0	0	0	0	0	0	0	0	0	0	1	1	0	0
अनिद्रा	0	1	0	1	1	0	1	1	1	0	0	1	2	1	0	0	0	1	1	0	0
ब्रोंकाइटिस	0	0	0	2	1	0	1	0	1	0	0	2	2	0	2	0	0	0	0	0	1
दुबलापन	0	0	0	0	0	0	0	0	0	0	0	0	0	0	0	0	0	0	0	0	0
मोतियाबिंद	0	0	0	0	0	0	0	0	0	0	0	0	0	0	0	0	0	1	0	0	0
बवासीर	0	0	1	0	0	1	1	0	0	0	0	0	0	0	0	1	0	0	0	1	0
बदहज़मी	0	0	1	0	0	0	0	0	0	0	0	0	0	0	0	0	0	1	0	1	0
अपच	0	0	3	2	0	0	0	1	1	1	1	1	1	0	0	0	0	2	0	3	0
अफारा	0	0	1	0	0	0	0	0	0	0	0	0	0	0	0	0	0	1	0	1	0
माइग्रेन	1	0	1	0	1	0	1	2	1	1	0	1	0	0	1	1	1	3	0	1	1
उबकाई	0	0	0	0	1	0	2	2	1	2	0	1	0	0	1	1	1	1	0	0	1
डीहाइड्रेशन	0	1	0	0	1	0	1	1	0	1	0	2	0	0	0	0	1	0	1	0	0
डायरिया	0	0	1	3	2	1	2	2	3	1	1	4	2	1	0	1	1	1	0	1	1
स्पोंडिलोसिस	0	1	0	0	1	0	0	0	0	0	0	1	0	0	0	0	0	0	1	0	0
थकान	2	8	4	10	22	3	8	1	1	2	1	11	7	2	3	4	1	8	8	4	1
स्पोंडिलोसिस	0	1	0	0	1	0	0	0	0	0	0	1	0	0	0	0	0	0	1	0	0
बेरीबेरी	0	1	0	0	0	0	0	0	0	0	0	0	0	0	0	0	0	0	1	0	0
अम्लपित्त	0	0	1	0	0	0	1	0	0	0	0	0	0	0	0	0	0	0	0	1	0

Fig. 2 Co-occurrence matrix for Hindi health dataset

4.3 Tolerant Pattern Learner (TPL) with Genetic Algorithm

We first begin the discussion by giving an overview of the TPL approach, followed by a brief introduction to the GA approach that we adopt in our TPL-GA algorithm. Readers can refer to [1, 2, 19] for details concerning formal and implementation aspects of TPL.

TPL Approach

TPL uses the tolerance approximation space to create an enriched representation for the promoted noun phrases. The noun phrases that are not yet categorized are described in terms of sets of co-occurring contextual patterns, rather than vectors. Here, noun phrases that are promoted are associated with three descriptive sets: $C(n_i)$, $U_A(n_i)$ and $L_A(n_i)$. These three values of the trusted instances are used to calculate a micro-score for a candidate noun that is not yet categorized. Equation 1 calculates the micro score for a candidate instance n_j against the trusted instance n_i is given below. The focus of this paper is on the hyper-parameters α, β and γ.

$$micro(n_i, n_j) = \omega(C(n_i), C(n_j))\alpha + \omega(U_A(n_i), C(n_j))\beta + \omega(L_A(n_i), C(n_j))\gamma \quad (1)$$

TPL uses a score-based sorting method using approximations. For each category *cat*, a macro score is calculated which is a summation of the micro scores of a particular instance.

$$macro_{cat}(n_j) = \sum_{\forall n_i \in cat} GA(micro(n_i, n_j)) \quad (2)$$

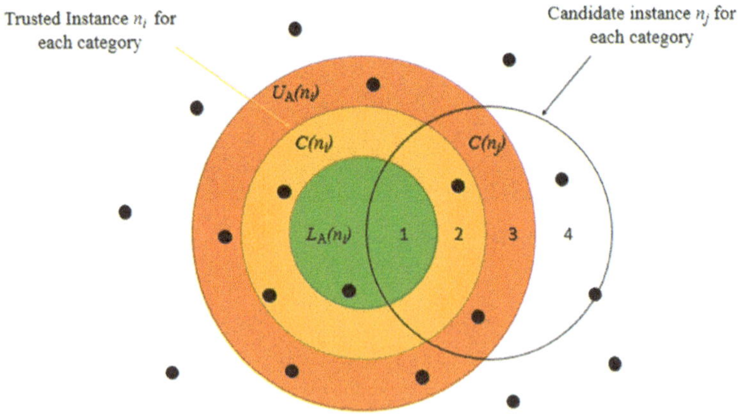

Fig. 3 Zone of recognition for context from approximations [1]

Figure 3 gives the basic principle of how the hyper parameters are affected by the various levels of approximation. For a trusted noun phrase n_i, its three descriptors partition the universe of contexts into four recognition zones. For calculating the micro score, a candidate noun's contexts falling in zone 1 of the trusted are covered by all three descriptors and thus will make a high contribution to its score. Similarly, context in zone 2 will make medium contribution and context in zone 3 will make low contribution. Whereas zone 4 does not make any contribution.

The candidates are sorted according to the normalized macro score based on the number of trusted instances belonging to that particular category. The TPL algorithm is designed to learn new instances at each iteration since the top scoring candidates are then promoted to trusted instances. There are two factors that impact the outcome of this algorithm. One is that the initial seeds for each category needs to be correctly mapped otherwise there could be false negative categorization. Another thing to note are the values α, β and γ while calculating the micro score. Currently, these values are chosen manually according to the domain or a particular dataset, hence these could be called as hyper-parameters. However, we propose the use of GA for these hyper-parameters. While applying genetic algorithms to the micro score hyper-parameters, the initial population is chosen at random from a fixed range. α, β and γ are the micro score parameters optimized using genetic algorithms.

Genetic Algorithm

Genetic Algorithm is based on the biological principle of "survival of the fittest" which ensures that the current population generates the best possible offspring [20]. It is a strategy-based programming technique which mimics the natural evolution process. The population consists of individuals where each individual represents a solution to the problem that is to be solved. Here, each individual is referred to as a chromosome. Chromosomes further consist of genes, where each elemental position in the chromosome is one gene. The algorithm starts with initial population values which are randomly generated. A fitness function is used to compare and find the best chromosomes from the population. New offsprings are generated by applying operations like crossover and mutation. These offsprings then replace some individuals from the initial population. This process is further repeated until we get the best possible solution [21]. Genetic Algorithms are both good and fast to produce an optimized result. This property makes it a good option for solving optimization problems. Therefore, we will be using GA to optimize the TPL algorithm for making it adaptive to text data of different domains and different languages.

TPL-GA Algorithm

The TPL-GA algorithm is given in Fig. 4 where line 8 is the focus of this work where the GA-based micro-scoring replaces the manual scoring used in TPL1.0 and TPL2.0 [19]. While implementing the genetic algorithm, one of the important decisions is to decide the representation of the chromosomes. Some of the most common representations are: binary, integer, real and permutation. In TPL-GA, we have used a real-valued representation since the values of α, β and γ used in the calculation of micro score of the TPL algorithm are real valued.

Algorithm: TPL with Genetic Algorithm

Input: Co-occurrence matrix and a small set of initial seed values, a Universal corpus U

Output: Trusted instances for each category

1. **for** $round = 1 \rightarrow \infty$ **do**
2. **for** each category *cat* **do**
3. **for** each new trusted member phrase n_i of *cat* **do**
4. Calculate the approximations $U_A(n_i)$ and $L_A(n_i)$;
5. **for** each candidate member phrase n_j **do**
6. Calculate $micro(n_i, n_j)$;
7. **for** each candidate member phrase n_j **do**
8. $macro_{cat}(n_j) = \sum_{\forall n_i \in cat} GA(micro(n_i, nj))$;
9. Rank instances by $macro_{cat}/|cat|$;
10. Promote top instances as trusted;

Fig. 4 TPL-GA algorithm

Figure 5 illustrates the overall flow of the GA component of the TPL-GA algorithm where the optimization is done on the micro score calculation, therefore we consider the micro-score as the fitness function (shown in Fig. 6).

For other GA operations, one-point crossover (shown in Fig. 7) is performed where one random combination point for chromosomes of both parents is selected in

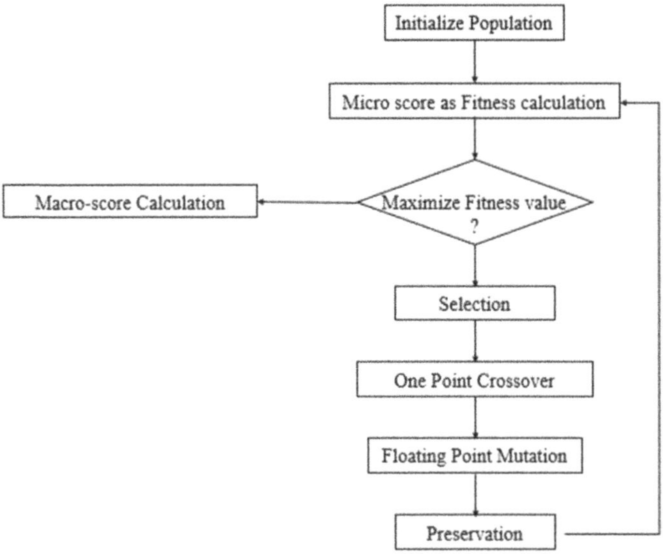

Fig. 5 Flowchart of genetic algorithm

```
Algorithm: Genetic Algorithm
Input: Micro-score values
Output: Optimized micro-score value
1. Random initialization of the population pop based on the domain of hyper-
parameter values.
2.  for each generation gen do
3.      fitness = Calculate_Fitness_Value(pop)
4.      parents = Selecting_Mating_Pool(fitness, pop)
5.      crossover = One_Point_Crossover(parents)
6.      mutation = Floating_Point_Mutation(crossover)
7.      new_pop = pop + mutation
8.  fitness = Calculate_Fitness_Value(new_pop)
9.  best_match = Sort(max(fitness))
```

Fig. 6 Pseudo code of genetic algorithm

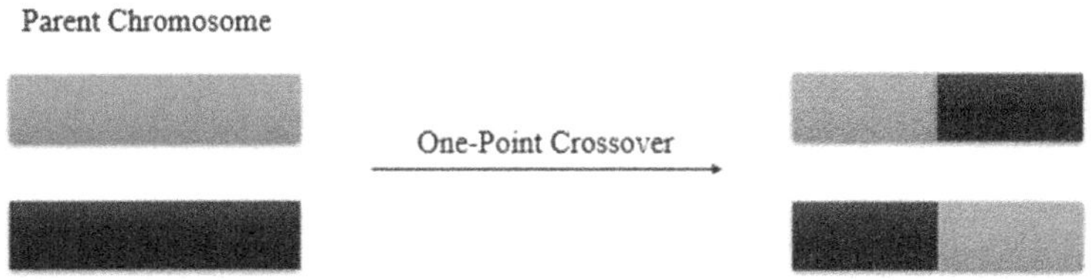

Fig. 7 Crossover operation

this operation. After these variations the chromosomal segment is switched together, giving birth to two new offspring following which floating point mutation is carried out to maintain genetic diversity from one generation of a population to the other. Finally, the best matched chromosomes are carried on to the next generation thereby, getting the best fit values.

For better results we have randomized the initial values of these hyperparameters between a certain range. There are different ways in which we can choose values of α, β and γ. Many such ways of choosing ranges of these hyperparameters and defining a relation between them were tested and the most suitable values were when $\alpha > \beta > \gamma$. Here α is assigned random values between [0.25, 0.75], β is roughly one-third the value of α and is assigned value between [0.25/3, 0.75/3] and finally γ is assigned between [0.25/9, 0.75/9]. However, the range of values for α, β and γ are configured according to the domain of the application.

4.4 Implementation Details

TPL-GA was implemented in Python 3 because of libraries such as Numpy which makes it easier to work with matrices. We used Ubuntu 16.04 LTS machine for the experimentation in Hindi text since Windows does not support encoding/decoding in Hindi language very easily. Since there was no existing work on the categorization of diseases in Hindi language, the list of symptoms and diseases were classified into 3 categories, Infectious, Hereditary and Lifestyle based disease. The generation of co-occurrence matrix which is used as an input to the optimized TPL algorithm was automated. To construct the matrix, various methods were tested to select all the possible pairs or words like using regex, selecting near-by box by choosing box of width 4 and identifying words in a particular sentence. As discussed earlier, the TPL algorithm uses three hyper-parameters known as weights which reflect the degree of similarity that is contributed by the entity belonging to that specific zone. Instead of manually choosing these hyper-parameters, we optimised it using the Genetic Algorithm. Various different ways of selecting the ranges and defining relations between these hyper-parameters were done. The optimized TPL-GA algorithm was tested on two different domains described in Sect. 4.1.

5 Results

We have performed a comparative analysis of TPL with TPL-GA using genetic operations for tuning. For all executions, we maintained a constant round of iterations. The assignment of the contributing factors for TPL is done randomly between the range [0.25, 0.75]. The hyper-parameter value assignment of TPL is as follows: $\alpha > \beta > \gamma$ (as shown in Fig. 3). This is because lower approximation has more contribution to the micro score value as compared to the candidate and the upper approximation. As a result, α is assigned initial random values between [0.25, 0.75], β is assigned between [0.25/3, 0.75/3] and finally γ is assigned between [0.25/9, 0.75/9].

The initial population size for the genetic algorithm is 20 and parent size is set to 2. The optimization is done on the micro score calculation, therefore it constitutes the fitness function. One-point crossover is performed following which mutation is carried out to maintain genetic diversity from one generation of a population to the other. Accuracy score is used as a performance metric for comparison. Accuracy was measured using manual inspection, where the entities in both the datasets were manually divided into categories to compare results between TPL and optimized TPL-GA. The results obtained are illustrated in Table 1.

For a trusted noun, the optimized TPL-GA gives higher macro scores for the candidate nouns as compared to the TPL algorithm which can be seen in Table 1. Based on the results shown in Tables 2 and 3, TPL algorithm with GA gives us a slightly better accuracy than TPL algorithm for all the three categories. Figure 8 and Table 4 shows the accuracy of TPL and TPL-GA on both the datasets.

Table 1 Top scores for n_i = Axe, England, Bread against n_j counterparts

Rank	Noun phrase n_i = Axe	TPL score	TPL-GA score	Noun phrase n_i = England	TPL score	TPL-GA score	Noun phrase n_i = Bread	TPL score	TPL-GA score
1	Axe	95.91	96.91	England	97.57	98.93	Bread	98.22	99.18
2	Ropes	94.89	94.29	India	9.56	97.59	Burger	94.73	98.91
3	Slings	90.25	92.13	KSA	91.54	97.59	Rice	90.84	97.23
4	Sword	86.59	89.87	Brazil	85.45	94.33	Salad	88.27	94.87
5	Fan	81.54	84.83	Afghanistan	81.79	91.48	Sausage	84.51	91.75
6	Whips	78.23	81.56	Argentina	76.23	87.56	Yogurt	79.56	87.99
7	Shield	69.56	72.23	USA	70.42	83.03	Seafood	74.38	82.93
8	Flamethrower	60.84	68.65	Sri-Lanka	63.51	79.93	Steak	69.87	72.49
9	Javelin	54.56	61.53	Pakistan	59.12	61.49	Ice-cream	61.54	63.33
10	Banton	49.56	55.26	South Africa	50.48	53.45	Egg	36.56	48.89

Table 2 Results on the Hindi health dataset for category infectious disease and hereditary disease

Iteration	Infectious		Hereditary	
	TPL	TPL+GA	TPL	TPL+GA
1	**डेंगू(Dengue)**	**डेंगू(Dengue)**	**दमा (Asthma)**	**दमा (Asthma)**
2	**मलेरिया (Malaria)**	**मलेरिया (Malaria)**	**कब्ज (Constipation)**	**कब्ज (Constipation)**
3	**एड्स (AIDS)**	**एड्स (AIDS)**	**मोतियाबिंद (Cataracts)**	**मोतियाबिंद(Cataracts)**
4	**खसरा(Measles)**	**खसरा(Measles)**	थकान (Fatigue)	थकान (Fatigue)
5	डायरिया (Diarrhea)	डायरिया (Diarrhea)	**मोटापा (Obesity)**	**मोटापा (Obesity)**
6	मधुमेह (Diabetes)	अपच (Indigestion)	माइग्रेन (Migraine)	माइग्रेन (Migraine)
7	अपच (Indigestion)	मधुमेह (Diabetes)	उबकाई (Emetic)	उबकाई (Emetic)
8	**हैपेटाइटिस(Hepatitis)**	**हैपेटाइटिस(Hepatitis)**	**ब्रोंकाइटिस (Bronchitis)**	**ब्रोंकाइटिस (Bronchitis)**
9	**निमोनिया(Pneumonia)**	**टाइफाइड(Typhoid)**	मेनिनजाइटिस (Meningitis)	मेनिनजाइटिस (Meningitis)
10	बवासीर(Piles)	**फाइलेरिया (Filariasis)**	अफारा (Afara)	**बवासीर (Piles)**
11	**टिटनस** (Tetanus)	**टिटनस (Tetanus)**	अम्लपित्त (Acidity)	अफारा (Afara)
12	**हैजा** (Cholera)	**हैजा (Cholera)**	गंजापन (Baldness)	गंजापन (Baldness)
Accuracy	66.6%	75%	50%	58.3%

Candidate diseases that are correctly classified are highlighted

Table 3 Results on the Hindi health dataset for category life-style based disease

Iteration	Life-Style based	
	TPL	TPL+GA
1	**बदहज़मी(Indigestion)**	**बदहज़मी (Indigestion)**
2	**डीहाइड्रेशन(Dehydration)**	**डीहाइड्रेशन (Dehydration)**
3	**थकान(Fatigue)**	**थकान (Fatigue)**
4	**स्पॉंडिलोसिस(Spondylosis)**	**स्पॉंडिलोसिस (Spondylosis)**
5	हाइपोथाइराइड (Hypothyroid)	हाइपोथाइराइड (Hypothyroid)
6	अनिद्रा (Insomnia)	अनिद्रा (Insomnia)
7	गठिया (Arthritis)	गठिया (Arthritis)
8	टाइफाइड (Typhoid)	निमोनिया (Pneumonia)
9	फाइलेरिया (Filariasis)	कुष्ठरोग (Leprosy)
10	कुष्ठरोग (Leprosy)	**अम्लपित्त (Acidity)**
11	**बेरीबेरी (Beriberi)**	**बेरीबेरी (Beriberi)**
12	दुबलापन (Thinness)	दुबलापन (Thinness)
Accuracy	41.6%	50%

Candidate diseases that are correctly classified are highlighted

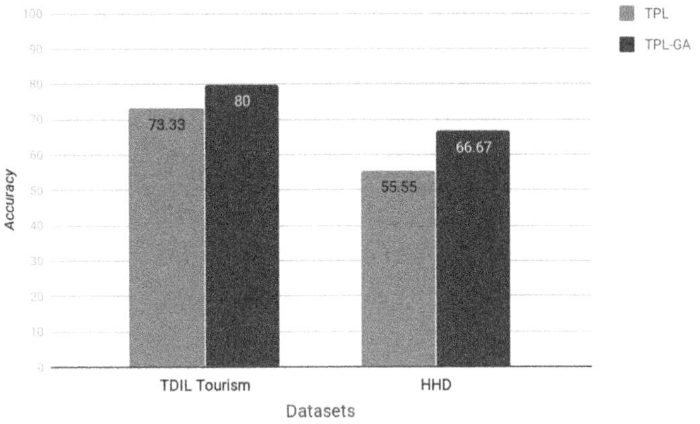

Fig. 8 Graphical representation of the accuracy metric

Table 4 TPL and TPL-GA accuracy results

Dataset	TPL accuracy	TPL-GA accuracy
TDIL tourism	73.33	80.00
HHD	55.55	66.67

6 Conclusion

In this research, the existing Tolerant Pattern Learner (TPL) algorithm for labelling nouns into different categories is improved by adding the Genetic algorithm (GA) feature. By introducing the TPL-GA on datasets from different vocabulary domains including the Hindi Health dataset, the overall performance is assessed. Starting with nouns and respective contexts from a few corpora, the co-occurrence matrix is built. Also, by incorporating the GA in the micro score calculation, an optimized macro score is computed, which results in the better ranking of instances—thereby resulting in trusted instances getting categorized within a small round of iterations. The quality of this matrix has a huge impact on the performance of the semi-supervised model. It is our observation that TPL-GA will perform better if more and conclusive seed examples are used. This is due to the fact that the scoring mechanism with hyper parameters will improve with a larger corpus, thus making it possible to identify the categories of the candidate nouns.

As future work, we plan to explore the capabilities of TPL over more categories and larger datasets for categorizing nouns and relations. We also intend to implement TPL on different linguistic datasets.

References

1. Sengoz, C., Ramanna, S.: Learning relational facts from the web: a tolerance rough set approach. Pattern Recogn. Lett. **67**(P2), 130–137 (2015)
2. Sengoz, C., Ramanna, S.: A semi-supervised learning algorithm for web information extraction with tolerance rough sets. In: AMT 2014, LNCS 8610. pp. 1–10 (2014)
3. Holland, J.: Adaptation in Natural and Artificial Systems. University of Michigan press, Ann Arbor (1975)
4. Ramanna, S., Peters, J., Sengoz, C.: Application of tolerance rough sets in structured and unstructured text categorization: a survey. In: Wang, G., et al. (eds.) Thriving Rough Sets, Studies in Computational Intelligence, vol. 708, pp. 119–173 (2017)
5. Pawlak, Z.: Rough sets. Int. J. Comput. Inform. Sci. **11**(5), 341–356 (1982)
6. Schroeder, M., Wright, M.: Tolerance and weak tolerance relations. J. Comb. Math. Comb. Comput. **11**, 123–160 (1992)
7. Mahdisoltani, F., Biega, J., Suchanek, F.M.: YAGO3: a knowledge base from multilingual wikipedias. In: 7th Biennial Conference on Innovative Data Systems Research (CIDR 2015)
8. Suchanek, F.M.: Automated construction and growth of a large ontology. PhD Thesis, Natural Sciences and Technology of Saarland University, 2009
9. Suchanek, F.M., Kasneci, G., Weikum, G., Yago, G.: A core of semantic knowledge. In: 16th International World Wide Web Conference (WWW 2007). ACM Press, New York, pp. 697–706 (2007)

10. Etzioni, O., Fader, A., Christensen, J., Soderland, S.: Open information extraction: the second generation. In: International Joint Conference on Artificial Intelligence, pp. 3–10 (2011)
11. Banko, M., Cafarella, M., Soderland, S., Broadhead, M., Etzioni, O.: Open information extraction from the web. In: Proceedings of IJCAI, pp. 2670–2676 (2007)
12. Carlson, A., Betteridge, J., Wang, R.C., Hruschka, E.R. Jr., Mitchell, T.M.: Coupled semi-supervised learning for information extraction. In: Proceedings of the 3rd ACM International Conference on Web Search and Data Mining, pp. 101–110 (2010)
13. Mitchell, T., Cohen, W., Hruschka, E., Talukdar, P., Betteridge, J., Carlson, A., Dalvi, B., Gardner, M., Kisiel, B., Krishnamurthy, J., Lao, N., Mazaitis, K., Mohamed, T., Nakashole, N., Platanios, E., Ritter, A., Samadi, M., Settles, B., Wang, R., Wijaya, D., Gupta, A., Chen, X., Saparov, A., Greaves, M., Welling, J.: Neverending learning. Commun. ACM **61**(5), 103–115 (2018)
14. Verma, S., Hruschka, E.R. Jr.: Coupled Bayesian sets algorithm for semisupervised learning and information extraction. In: ECML PKDD Part II LNCS 7524, pp. 307–322 (2012)
15. Bharadwaj, A., Ramanna, S.: Categorizing relational facts from the web with fuzzy rough sets. Knowl. Inf. Syst. **61**, 1695–1713 (2019)
16. Razali, N.M., Geraghty, J.: A genetic algorithm performance with different selection strategies. In: Proceedings of the World Congress on Engineering, vol. II (2011)
17. Jain, A., Arora, A.: Named entity recognition in Hindi using hyperspace analogue to language and conditional random field. Pertanika J. Sci. Technol. UPM **26**(4), 1801–1822 (2018)
18. Jain, A., Tayal, D.K., Arora, A.: OntoHindi NER—an ontology based novel approach for Hindi named entity recognition. Int. J. Artif. Intell. **16**(2), 1–36 (2018)
19. Ramanna, M.: Patterns **1**, 100053 (2020). https://doi.org/10.1016/j.patter.2020.100053
20. Thengade, A., Dondal, R.: Genetic algorithm—survey paper. In: IJCA Proceedings National Conference on Recent Trends in Computing, NCRTC, 5 (2012)
21. Radwan, A., Latef, B., Ali, A., Sadek, O.: Using genetic algorithm to improve information retrieval systems. World Acad. Sci. Eng. Technol. **17**(2), 6–13 (2006)

Data-Driven Fuzzy C-Means Equivalent Turbine-Governor for Power System Frequency Response

Jose A. Barrios-Gomez, Francisco Sanchez, Francisco Gonzalez-Longatt, Gianfranco Claudio, Alberto Cavazos, Harold R. Chamorro, and Wilmar Martinez

Abstract This research paper proposes a turbine-governor modelling technique based on equivalent FCM (Fuzzy C-Means) for a control area of an equivalent power system used for frequency response analysis. The FCM algorithm implementation is proposed to find an equivalent Fuzzy model of n turbine-governors that are in an area of the electric power system (EPS). The FCM algorithm is mainly used to generate the rules for the fuzzy model; this algorithm uses input-output data, deviation of frequency, velocity and its derivatives, these are numerical data of a control area of the electrical system that contains n turbine-governors. Two cases are used to test the equivalent FCM model: (i) model of three areas simulink model, where area two has been modified by adding four turbine-governors to verify that it is possible to define an equivalent FIS model based on data, and (ii) the multi-area system, that is extracted from a reduced system frequency response model of an electric area of Great Britain Power System (GBPS), which contains three different types of turbine-

J. A. Barrios-Gomez · F. Sanchez · G. Claudio
Centre for Renewable Energy Systems Technology, CREST Loughborough University, Loughborough, UK
e-mail: J.A.Barrios-Gomez@lboro.ac.uk

F. Sanchez
e-mail: F.Sanchez@lboro.ac.uk

G. Claudio
e-mail: G.Claudio@lboro.ac.uk

J. A. Barrios-Gomez · A. Cavazos
Facultad de Ingeniería Mecánica y Eléctrica, Universidad Autónoma de Nuevo León, San Nicolás de los Garza, 66451 Nuevo León, Mexico
e-mail: alberto.cavazosgz@uanl.edu.mx

F. Gonzalez-Longatt
University of South-Eastern Norway, Porsgrunn, Norway
e-mail: f.gonzalez-longatt@usn.no

H. R. Chamorro (✉) · W. Martinez
Department of Electrical Engineering (ESAT), KU Leuven, Diepenbeek, Belgium
e-mail: hr.chamo@ieee.org

W. Martinez
e-mail: wilmar.martinez@kuleuven.be

D. Gupta et al. (eds.), *Soft Computing for Data Analytics, Classification Model, and Control*, Studies in Fuzziness and Soft Computing 413,
https://doi.org/10.1007/978-3-030-92026-5_7

governors, the data for this model was obtained from DIgSILENT-PowerFactory. The equivalent fuzzy model is tested under the same conditions as the original system with n turbine-governors, and they are compared against each other. The simulation results and performance analysis show it is possible to find an equivalent model with excellent performance with FCM and that the parameters of the FIS model can be adjusted if necessary, with ANFIS.

Keywords FCM systems · Fuzzy logic systems · Fuzzy modelling · Intelligent systems · Power systems frequency response

1 Introduction

Nowadays, the electrical industry is continuously looking for ways to improve the efficiency and reliability of energy production. It is well-known that turbine-governor is a device that has an essential task to keep the frequency constant. In most cases, the EPSs are constituted by several control areas, within which it is possible to find different types of power generation such as hydroelectric, coal, gas, biomass, etc. for example the Great Britain Power System (GBPS) in Fig. 1.

The modelling of the turbine-governor is of great importance when studying frequency control and stability. Frequency response is an essential aspect of a power system performance [1]. In the literature, it is possible to find models of turbine-governors designed based on human knowledge and some modelling techniques with artificial intelligence such as fuzzy logic, neural networks and genetic algorithms.

The turbine-governor modelling is essential for the stability studies of the PS deal primarily with transient angle and voltage stability, small-signal stability and frequency control and stability. The definitions of each of these types of stability problems can be seen in [2].

Typically, it is not feasible to separate the various stability problems, in small systems, a single event may lead to multiple interrelated phenomena of frequency, voltage and angular stability problems. The critical point of the dynamics of the turbine-governor is the initial response of the turbine-governor in the first seconds after a disturbance of the network [3].

Different techniques have been applied to model the turbine-governor controller; some are PID controllers, intelligent controllers and adaptive controller [4–8].

However, traditional PID controllers do not respond efficiently when there are non-linearities in complex processes. This is where fuzzy logic becomes an important and successful technology, which contributes with excellent results in the control and modelling of the EPS and is used in various areas, primarily for non-linearities. They have also been used very successfully in almost all technical fields such as expert systems, image and signal processing, control and modelling [9–12]. A fuzzy-based turbine-governor model was proposed using human experience and recognise the importance it has for maintaining constant frequency. A mathematical model was used to simulate the hydroelectric power station; simulation models are suitable for

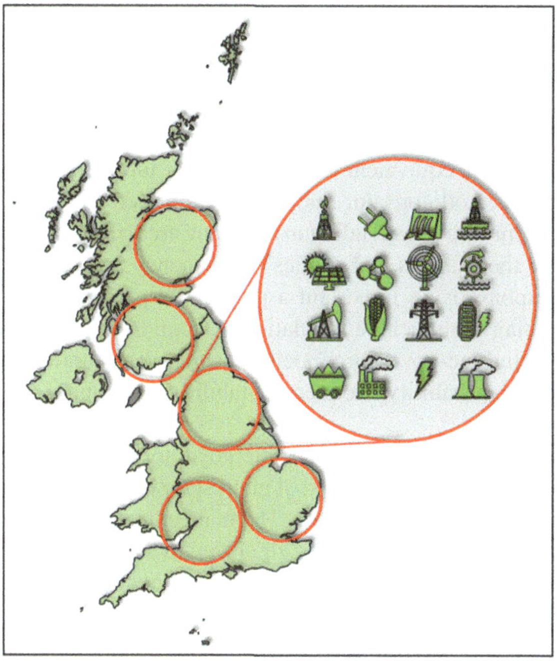

Fig. 1 Different types of generation to the GBPS

use in large-scale systems stability studies. It was shown that the fuzzy controller has speed and robustness and can maintain the characteristics of the electrical energy generated despite the changing load [13].

There has also been the design of governors applying other artificial intelligence techniques, in [14] presents the design of the PID controller based on the artificial neural network (ANN), to perform a rapid action of the regulator in a power generation plant. A model based on an artificial neural network of the governor-turbine system using measured test data from the governor. The dynamics of the proposed governor-turbine system are modelled quite accurately, and they conducted comparative studies between the proposed and conventional models [14].

Research and studies have also been done for modelling the turbine-governor: A linear and nonlinear mathematical model and design of Hydraulic turbine-governor system is proposed in [15], the design of the hydraulic turbine regulator is performed using proportional control with a constant and transient drop, integral proportional control (PI) and Proportional Derivative Integral control (PID). A Modeling of the theoretical steam turbine-governor and parameter tests for simulation of the PS is proposed in [16]. The theoretical modelling, parameter testing and model correction for a steam turbine regulator are also discussed.

An equivalent use delay model is used to calculate an equivalent turbine-governor model, this article modelling the effects of turbine-governor dynamics on the average frequency behaviour of a multi-generation system after a large generation loss or load change when the system remains synchronised [17]. To study the dynamics of frequency, reduced or equivalent models have been used as the model Average System Frequency (ASF). A frequency response model is reduced to the ASF model from which a uniform frequency can be achieved. The ASF model can be modelled with differential-algebraic equations and resolved with step-by-step integration [18].

In this investigation, a data-based Fuzzy C-Means (FCM) approach is proposed to find an equivalent model of a set of turbine-governors in parallel. Also describes the methodology for finding the simplified equivalent model of a turbine-governor assembly corresponding to a specific area of an EPS since this is useful and important for the analysis and studies of frequency control and stability in PS, the frequency response is an important aspect of PS performance, this analysis is useful to pre-tuning the machine regulation system, as mentioned in [19]. It is noteworthy that the turbine-governor modelling is of great importance for studies related to transient angular stability, frequency stability and control, to a lesser extent, small–signal stability.

Therefore, different turbine-governor blocks are used, the analysis of these components is important, and of interest for the simulation of electrical systems, a method for modelling is proposed. The main contribution of this document is the application of an FCM data clustering algorithm to find an equivalent model of turbine-governors set located in an electric area, the parameters of this diffuse model are optimised by applying ANFIS algorithm.

The rest of this chapter is organized as follows: Sect. 2 describes the governors frequency response preliminaries. Section 3 explains the methodology used for the FCM and the data required. Section 4 presents the ANFIS modelling. Section 5 introduces the test systems studied. Section 6 shows the obtained ANFIS turbine-governor equivalent. Section 7 presents the results obtained. Finally, the conclusion are given in Sect. 8.

2 Power Systems Preliminaries

Normally, when there is an isolated electrical area containing a generating unit or group of units, it is positioned nearby to distribute electricity in the same area, this is called a single area system. Only generating units present in that area are responsible for maintaining the desired frequency under normal and abnormal conditions [20]. The generator-load dynamic relationship, between the incremental mismatch power $P_m - P_L$ and the frequency deviation ω can be expressed as:

$$P_L - P_m = 2H\frac{d\omega}{dt} + D\omega \tag{1}$$

where ΔP_m is the mechanical power change, ω is the frequency deviation, ΔP_L is the load change, D is the load damping coefficient, and M is the inertia constant.

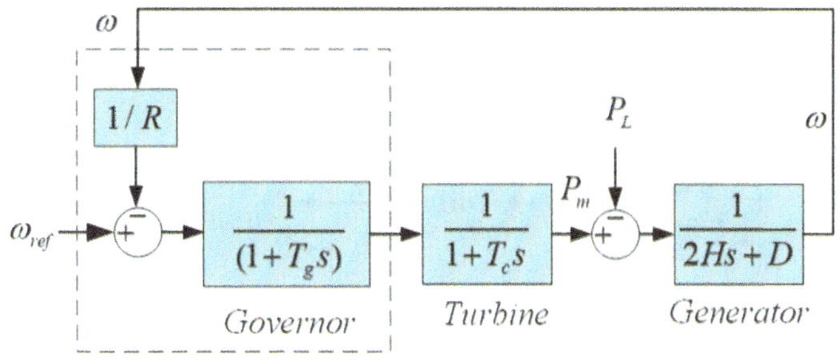

Fig. 2 Steam turbine in power single area

Low-order models have been proposed for the representation of the dynamics of turbines and generators for the application in the frequency analysis of power systems and the control design. The slow and fast dynamics of boiler and generator systems are generally excluded in these models [21]. Figure 2 shows the block diagram representation of the speed controller and the turbine for the steam and hydraulic regulating units suitable for load frequency control analysis, where R is the speed characteristic [20, 22].

The supplementary loop carries out feedback via the frequency deviation and adds it to the primary control loop through a dynamic controller. The signal obtained P_C is used to regulate the system frequency; in real PS, the dynamic controller is usually a simple integral or Proportional-Integral (PI) controller. In Fig. 2 it is shown, how the frequency experiences a transient change ω following a change in load P_L. Therefore, the feedback mechanism comes into play and generates an appropriate signal for the turbine to make generation P_m track the load and restore the system frequency [22].

The turbine-governor model is normally linked to one or two synchronous generators and is used to determine the mechanical power of the shaft (P_m) or torque for the generator model. A block diagram showing the relationship of the turbine-governor with the EPS is presented in Fig. 3.

The conventional automatic generation control (AGC) is used to achieve the secondary frequency control. In order to regain the initial operation frequency, the speed-droop curve must be shifted. This is done by changing the reference point of the produced power, ω_{ref}, for the generating units governing systems [23].

The governor has in the input speed, and the output controls the gate of the turbine, so the output of the turbine has mechanical power or torque [24]. Different energy sources determine the physical characteristics and categories of hydraulic turbines and governors, so each type requires its unique modelling. The delays in the hydraulic turbine and governor models are due to the compensation of transient drop, pilot valves and gate servomotors. Also, the effect of water inertia on hydraulic turbines significantly influences the governor's requirements and must be compensated with speed feedback and should be reflected in the modelling of these devices [25].

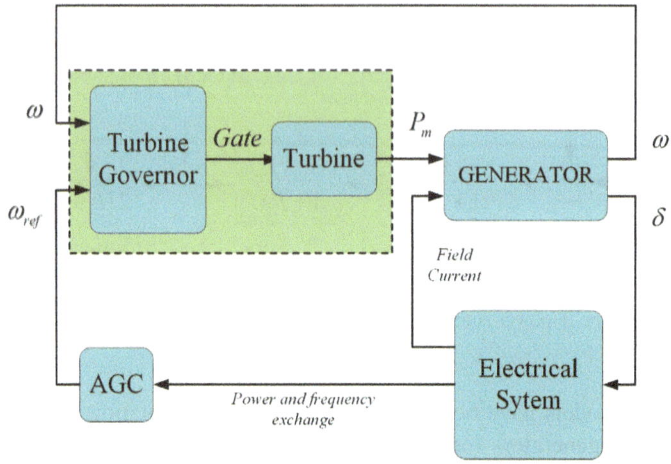

Fig. 3 Relationship of the turbine-governor with the overall power system

3 Methodology

In this research work, an equivalent model of n electrical regulators is developed, where n is the number of turbine-governors in the electrical area, applying FCM algorithms based on data-fuzzy modelling is performed, with the aim of establishing a methodology in the FCM application and optimising its parameters with ANFIS to find an equivalent turbine-governor. Since the modelling is based on data, which means that a series of numerical input-output data will be used to generate a rule base and with these create a fuzzy model. The data for the creation of the diffuse model is obtained from a pair of PSs to simulate two cases; the first is a simulated multiple areas system in MATLAB and the second is a simulated GBPS in the DIgSILENT-PowerFactory software, described further in Sect. 4.

In order to find an equivalent model, the following methodology was performed:

1. The input-output variables of the set of n turbine-governors of an EPS are selected, this will provide the corresponding input-output data set, it usually is the sum of ω and ω_{ref}, as seen in Fig. 3. Thus, there will be a quantity of data, which will depend on the selected sampling time and period.
2. Using the FCM algorithm with the desired inputs and outputs and through iterations regarding the data, the centroids or centres of mass are found and assign each data to the nearest centroid, based on an optimisation criterion, the algorithm groups similar data. Clusters are generated where the data does not necessarily belong to a single group, since it is a fuzzy modeling each data could belong to one or more groups, because FCM assigns to each data a membership value within each group, thus a data it may belong partially to more than one group. This solves the problem that occurs in the case that one of the data is close enough to two

groups in such a way that it is difficult to label it in one or the other. This FCM algorithm performs a smooth partition of the data, unlike the classic K-means algorithms that work with a hard partition.

3. The type of fuzzy inference to use is chosen: Sugeno or Mamdani; if it is necessary to adapt parameters, Sugeno models are used, only ANFIS allows this type of FIS. A quantity number of groups must be assigned so that the FCM system finds the centres of each of them. A small number of groups means that there will be fewer calculations to find the centroids of each group; a larger number will indicate otherwise.
4. A FIS is generated using the fuzzy grouping FCM by extracting a set of rules that models the behaviour of the data, as mentioned before FCM is used to determine the initial sets of the rules of an FIS. The rule extraction method first uses the FCM function to determine the number of membership rules and functions for the antecedent and consequent. The number of clusters determines the number of membership rules and functions in the generated FIS.
5. The model found with FCM is evaluated with a set of test data, and the error is calculated to analyse the performance of the FCM model.
6. After analysing the performance error of the FCM model, if it is determined that it is very large and does not meet the established margin of error. This can be improved using ANFIS. This algorithm will be trained based on the model created by FCM and a set of input-output training data, which will generate a new diffused set of trained FIS. Different numbers of times or iterations can be defined to reduce the error; this will depend on how complicated the model can behave.

3.1 FCM Modelling Systems

In this section, the algorithm used is briefly explained. FCM is a data cluster algorithm which partitions input-output data pairs into N groups or cluster that allows each data point to belong to multiple clusters with varying degrees of membership according to some similar characteristics given as an objective function which has to be minimised [26]. Such clustering algorithms are commonly used to determine the initial set of rules of an FIS. A collection of n vectors x_j, $j = 1, 2, 3, \ldots, n$, is partitions in to c groups C_i, $i = 1, 2, 3, \ldots, c$; c clusters centers are then determined J_m. FCM performs the following process to perform the clustering: Cluster membership data are randomly initialised, μ_{ij}, then calculate the centre of each group.

$$c_j = \frac{\sum_{i=1}^{d_p} \mu_{ij}^m x_i}{\sum_{i=1}^{d_p} \mu_{ij}^m} \tag{2}$$

μ_{ij} is updated from the equation:

$$\mu_{ij} = \frac{1}{\sum_{k=1}^{N} \left(\frac{\|x_i - c_j\|}{\|x_i - c_k\|} \right)^{\frac{2}{m-1}}} \tag{3}$$

By iteratively updating the cluster centres and the membership grades for each data point, FCM moves the cluster centres to the correct location within a data set and, for each data point, finds the degree of membership in each cluster and calculating the objective function through the equation

$$J_m = \sum_{i=1}^{d_p} \sum_{j=1}^{N} \mu_{ij}^m \left\| x_i - c_j \right\|^2 \tag{4}$$

where N is the number of clusters, d_p is the number of data points, x_i is the ith data point.

- c_j is the center of the jth cluster.
- m is fuzzy partition matrix exponent for controlling the degree of fuzzy overlap, with $m > 1$. Fuzzy overlap refers to how fuzzy the boundaries between clusters are, that is the number of data points that have significant membership in more than one cluster.
- μ_{ij} is the degree of membership of x_i in the jth cluster. For a given data point, xi, the sum of the membership values for all clusters is one [27].

3.2 *Fuzzy Logic and Fuzzy Inference System (FIS)*

Fuzzy Logic is basically a multivalued logic is an extension of classical boolean logic, where there is only "true" and "false" and assigns deterministic values to its variable [28]. Whereby a large part of "natural" reasoning is satisfied, but human reasoning uses "true" or "false" values, which are not "so deterministic." fuzzy logic aims to produce accurate results from inaccurate information [29].

The design of fuzzy systems depends in most cases the experience of a human expert. Fuzzy modeling provides a formal methodology to represent, manipulate and implement a human heuristic knowledge on how to control a system [30]. On the other hand, when working with more complex models, there are also techniques to design a diffuse model based on numerical data from the system [31]. During the fuzzy inference process the mapping of an input given to an output using fuzzy logic is proposed, this provides a basis from which decisions can be made. This process includes membership functions, fuzzy logic operators, IF THEN type rules and knowledge base, consisting of four interfaces:

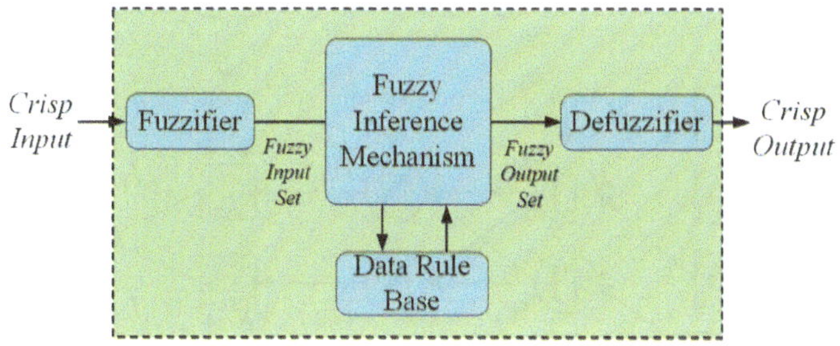

Fig. 4 Fuzzy logic system architecture

1. Fuzzifier.
2. The Rule-Base.
3. Inference mechanism.
4. Defuzzifier.

The fuzzifier interface modifies the inputs so that they can be interpreted and compared to the rules in the rules base. The "rule base" contains the knowledge, in the form of a set of rules, of the best way to control the system. The inference mechanism evaluates which control rules are relevant at the present time and then decides what the entrance to the plant should be [32]. The defuzzifier interface converts the conclusions reached by the inference mechanism at the input to the plant [30, 33]. Figure 4 shows the block diagram of a fuzzy system architecture.

4 ANFIS Modelling

The neuro-fuzzy adaptive inference system (ANFIS) were proposed by [33], it is a neuro-fuzzy hybrid learning adaptation algorithm, which is functionally equivalent to the Sugeno inference system [33, 34]. In this algorithm the individual outputs of each rule in FIS Sugeno are obtained and the output of the ANFIS system is obtained through an objective function, where the weights $\bar{w}_1$ and $\bar{w}_2$ are the normalised values of w_1 and w_2 with respect to the sum $(w_1 + w_2)$. The Sugeno equivalent ANFIS architecture mechanism and the output of the 2-rule ANFIS system is show in Fig. 5, is obtained by:

$$f = \sum_i \bar{w}_i f_i = \bar{w}_1 f_1 + \bar{w}_2 f_2 \tag{5}$$

Since the inference mechanism of the ANFIS system must be of the Sugeno type, the Sugeno systems created by FCM can be used for ANFIS by the MATLAB toolbox. A training data set is used to train the ANFIS model while the validation set was used to validate it [29].

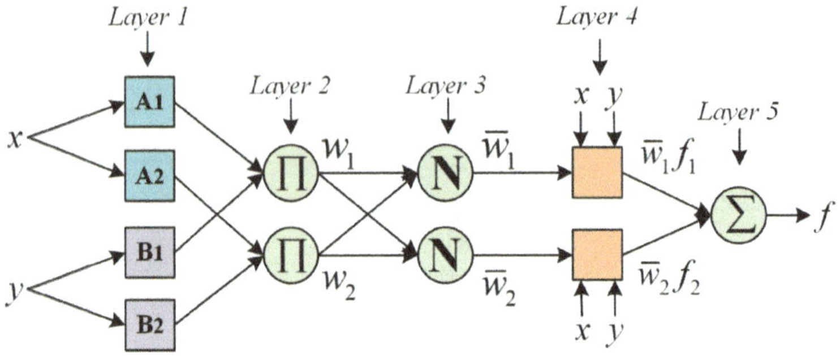

Fig. 5 ANFIS equivalent architecture

5 Test Power Systems

To demonstrate the methodology, two cases are proposed, the first is a model of three areas, where one of the areas has been modified by adding four governors, as shown in Fig. 6, these changes were made with the purpose of verifying that it is possible define an equivalent FIS model based on data. The second case is extracted from an electric area of the GBPS, which contains three turbine-governors, as seen in Fig. 7. The way in which the input-output data sets are taken in each case, due to the same conditions of each model, will test the methodology and performance of it.

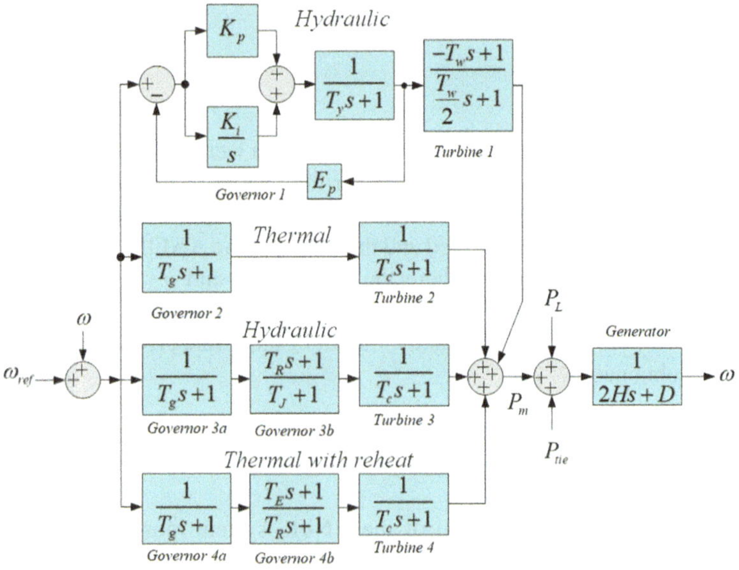

Fig. 6 Modified model of the three-area power system

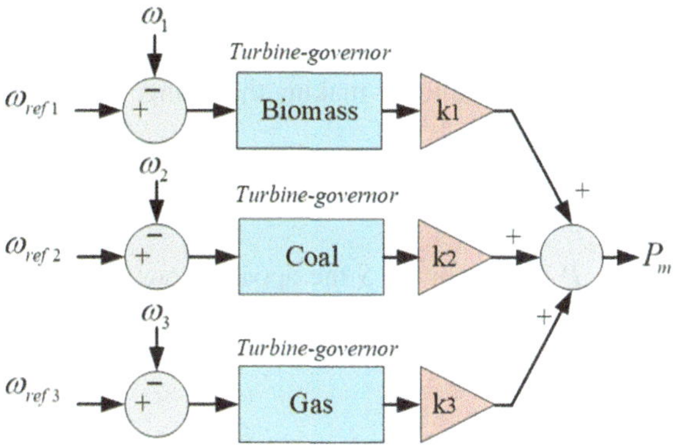

Fig. 7 Part of an area of the GBPS

5.1 *Test System I: Modified Three Area System*

A multi-area PS comprises areas that are interconnected by high-voltage transmission lines or link lines. The trend of the frequency measured in each control area is an indicator of the tendency of the mismatch power in the interconnection and not only in the control area. The Load-Frequency Control (LFC) system in each control area of an interconnected (multi-area) energy system must control the exchange energy with the other control areas, as well as its local frequency.

The proposed three-areas system consists of a model of three electric areas where each area contains a turbine-governor assembly. In this model area two has been modified as shown in Fig. 6, where is rotor speed and is reference speed, in order to simulate different types of power generation, four types of turbine-governor are used in this case in order to find an equivalent model with FCM.

5.2 *Test System 2: Great Britain Power System*

Most developed countries consider carbon dioxide (CO_2) emissions a high priority. They direct their efforts to make their EPS more environmentally friendly. The UK government is within these countries and aims for 2020 that 15% of energy generation comes from renewable energy sources and want to reduce CO_2 emissions by 14% of national transport could be achieved by supporting electric vehicles trials. The variability of renewable energy generation causes uncertainties, and this raises inevitable concerns about the operation of the electrical system [35]. A high penetration of renewable energy, in particular wind energy, will increase the difficulty of frequency regulation, several studies have been conducted to investigate the frequency control with high penetration of these types of energy [36, 37].

GBPS consists of several interconnected control areas that contain different types of power generation as shown in the study case where three types of generation, coal, biomass gas are taken. In Fig. 7, it is possible to see an extract of three sets of generation each with the corresponding turbine-governor block, these have as input the sum of the rotor speed and the reference speed, at the output we can see a gain k_i that is responsible for making the change from base to pu equation (6).

$$k_i = \frac{P_{Ni}}{S_{Ni}\cos(\phi)} \tag{6}$$

where, P is the power, S the apparent power and $\cos\phi$ is the angle, the output of these three sets of turbine-governor gives mechanical power output.

6 FCM and ANFIS Based Equivalent Turbine-Governor

Previously, the fundamentals of the theories of Fuzzy Logic, FCM, ANFIS were described, which were applied to be able to model the turbine-governor of the two proposed case studies. In Fig. 8, it can be seen that the speed difference of the measurement system is taken as an input for FIS, also the rotor speed is compared with the reference speed to find the speed deviation. The FIS output control signal is the control signal, which is the mechanical power, which will go to the generator block. The equivalent system based on suggested fuzzy is shown in Fig. 8.

The speed difference and its rate of change are the input to the FIS. The fuzzy system proposed as an equivalent turbine-governor is shown in Fig. 9. The frequency error and its derivative are the inputs to the FIS model.

The equivalent system based on FCM are Sugeno type, in Fig. 10 the input one (in1) and Fig. 11 the input two (in2) shows the fuzzy input sets with the corresponding membership functions that were found for the three-areas model. For such electrical system, tests were carried out which it was determined that 10 clusters were enough for the FCM algorithm to find an equivalent turbine-governor model.

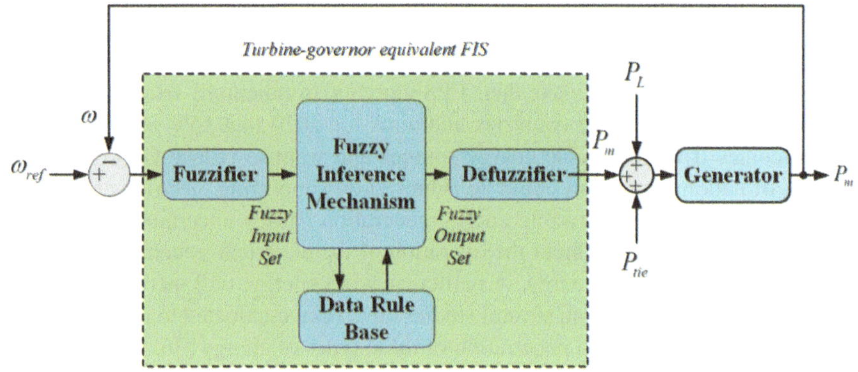

Fig. 8 Equivalent FIS structure

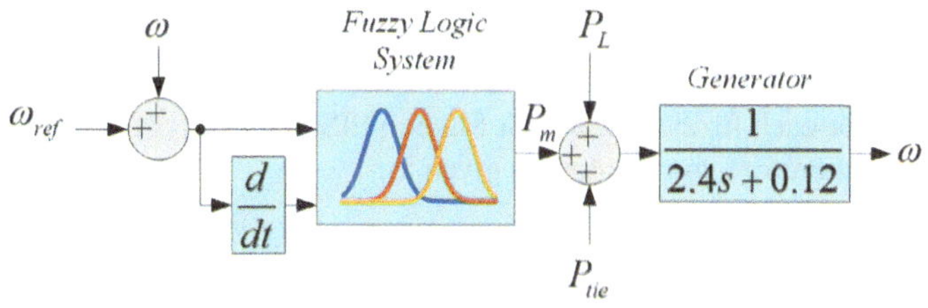

Fig. 9 Equivalent fuzzy based turbine-governor

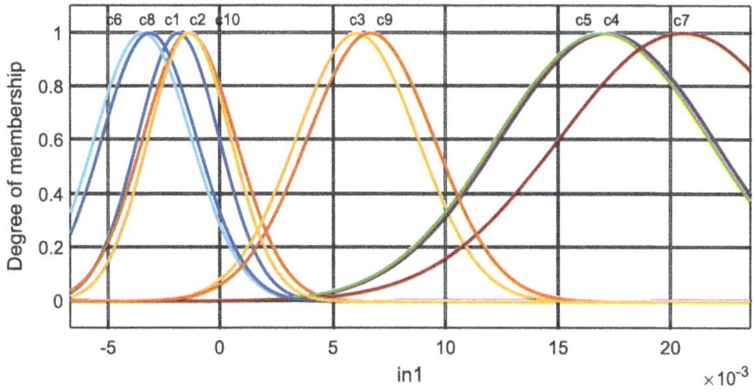

Fig. 10 Error membership functions

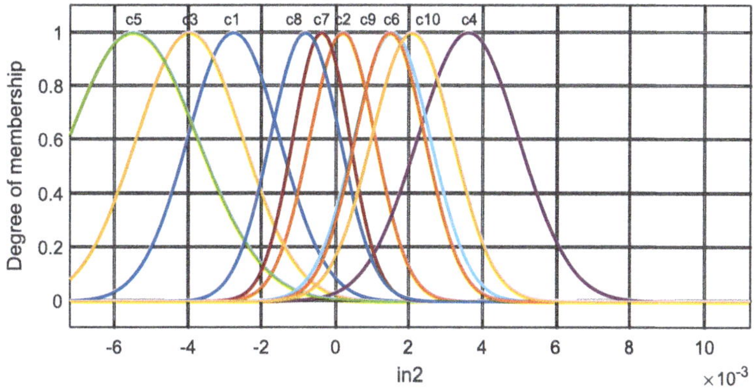

Fig. 11 Error rate of change membership functions

In this model, each input variable has a "gaussian" input membership function for each fuzzy cluster and each output variable has one output membership function type "linear" for each fuzzy cluster. For Sugeno FIS systems type, there is no fuzzy set at the exit, therefore since output is given as:

$$\text{output} = \frac{\sum_i w_i p_i}{\sum_i w_i} \tag{7}$$

where, w_i is the firing strength of the ith rule of the equivalent system, and p_i is given by the mean value of the ith output membership function for each fuzzy cluster [38].

The FIS generated for the equivalent turbine-governor contains one rule for each fuzzy cluster, for this system it will have ten cluster, therefore there will be ten rules. For example, a rule is:

"if in1 is cluster 1 AND in2 is cluster 1 THEN out is cluster 1"

In Fig. 12, it shows the fuzzy rules for the equivalent turbine-governor model of the three-areas electrical system case.

In the case of the GBPS, this system has three different types of turbine-governor, for this case it is considered that it will have as input and its derivative for each turbine-governor, in total six inputs, unlike from the previous case where he only had two. For the GBPS, tests were carried out, it was determined that it was necessary to normalise the data between 0 and 1, it was also tested with different number of clusters, with which 100 clusters were defined for the FCM algorithm to find a turbine-governor model of equivalent turbine, however it was necessary to adjust the parameters of the FCM model using ANFIS, in order to obtain a more accurate model. As in the case of the three-areas model, the FIS generated for the equivalent turbine-governor contains one rule for each fuzzy cluster, this system has 100 clusters, so it has 100 rules. An example of these rules is as follows:

	in1 is		**in2 is**		**out is**
IF	cluster 1	**AND**	cluster 1	**THEN**	cluster 1
	cluster 2		cluster 2		cluster 2
	cluster 3		cluster 3		cluster 3
	cluster 4		cluster 4		cluster 4
	cluster 5		cluster 5		cluster 5
	cluster 6		cluster 6		cluster 6
	cluster 7		cluster 7		cluster 7
	cluster 8		cluster 8		cluster 8
	cluster 9		cluster 9		cluster 9
	cluster 10		cluster 10		cluster 10

Fig. 12 FIS rule for equivalent turbine-governor three-areas system

	in1 is		in2 is	…	in8 is		out is
	c1		c1		c1		cluster 1
	c2		c2		c2		cluster 2
			⋮		⋮		⋮
IF	⋮	AND	⋮	…	⋮	THEN	⋮
	c98		c98		c98		cluster 98
	c99		c99		c99		cluster 99
	c100		c100		c100		cluster 100

Fig. 13 FIS rule for the equivalent turbine-governor of the GBPS

In Fig. 13, it shows a summary of the fuzzy rules for the equivalent turbine-governor model of the GBPS case. In this figure, the word cluster is abbreviated with the letter *c*.

7 Simulation Results

This section presents the results that were obtained from the simulation of the equivalent models for the two proposed cases. Figure 14 shows graphs comparing the mechanical power of the three-areas PS against the model found by FCM. In Fig. 15

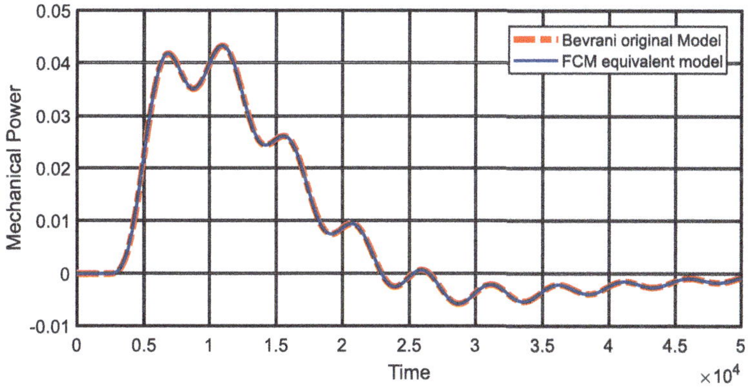

Fig. 14 Mechanical power for the three-area PS

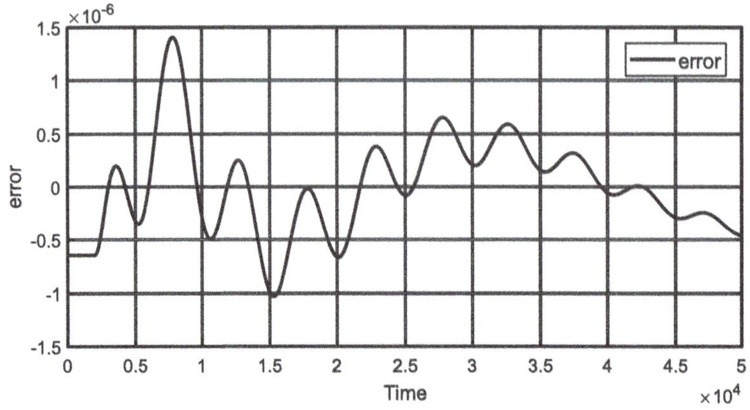

Fig. 15 Estimation error of the original three-areas model against the equivalent FCM model

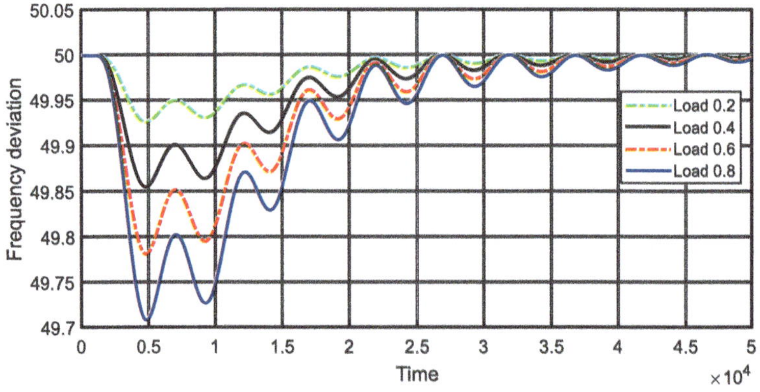

Fig. 16 Frequency deviation with different loads

shows the estimation error of the original three-areas model against the equivalent FCM model. The average estimation error obtained for this system is:

In addition, an extra test has been carried out to demonstrate that the FCM model works when the load is changed in one of the three-areas electrical system, tests were made for four different loads as can be seen in Fig. 16.

The second case, an area chosen to test the GBPS, considering that the data was simulated and obtained in DIgSILENT, which is a very realistic simulation tool, it was found that this GBPS has more complex behavior, this made it difficult to obtain a equivalent model turbine-governor, so it was necessary to use ANFIS, as mentioned above, the final result can be seen in Fig. 17.

In Fig. 18 shows the estimation error of the original three-areas model against the equivalent FCM model. The average estimation error obtained for GBPS is:

$$\text{error} = \text{FCM model} - \text{Original model mean(error)} = 5.3914\text{e} - 11.$$

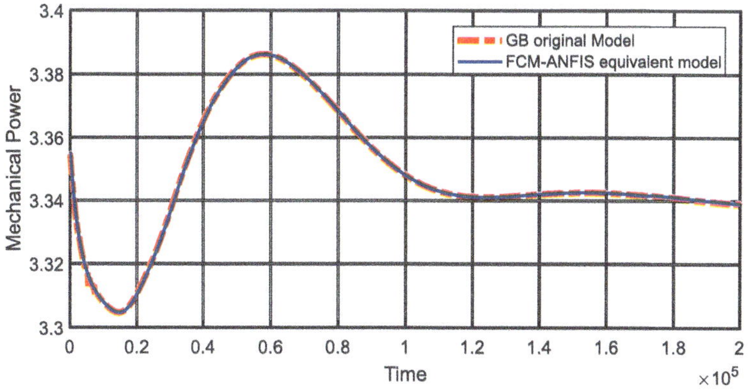

Fig. 17 Frequency deviation with different loads

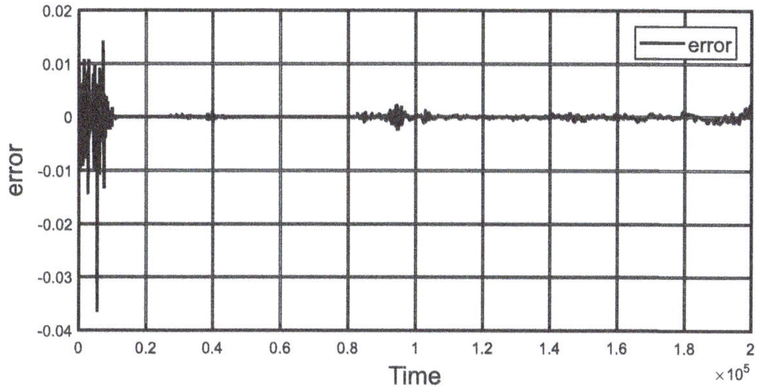

Fig. 18 Estimation error of the original GBPS against the equivalent FCM model

8 Conclusions

The implementation of modeling by FCM and ANFIS in EPSs is considered of great importance, since it provides the benefits of artificial intelligence in solving problems with high uncertainty. A turbine-governor FIS base was proposed. From the graphical results of the simulation and average estimation error that was very low, it was possible to demonstrate that the turbine-governor model based on FCM has excellent performance. It keeps the frequency constant despite changing the user's load at any operational point. Fuzzy controllers are considered suitable for implementation due to the nonlinear behavior of synchronous generators.

Two cases were proposed to demonstrate this, the three-area electrical system was only necessary 10 rules to find an equivalent model with low accuracy error. A test was performed using the model found to simulate load changes in the three-area system, behaved as expected. For the second case, the GBPS presented a more complicated behavior, in addition to using an FCM system of 100 clusters, it was necessary to adjust the model with ANFIS, ten training epochs were necessary to have a good performance.

Acknowledgements The author is thankful to CONACYT for the financial assistance in this work at the Loughborough University, UK. Jose Angel Barrios Gomez is grateful to CONACYT-Mexico for providing a postdoctoral research fellowship.

References

1. Crary, S.B., McClure, J.B.: Supplementary control of prime-mover speed governors. Trans. Am. Inst. Electr. Eng. **61**(4), 209–214 (1942)
2. Ruud, H.B., Farnham, S.B.: A new automatic load control for turbine generators. Trans. Am. Inst. Electr. Eng. **68**(2), 1337–1342 (1949)
3. Pourbeik, P., Chown, G., Feltes, J., Modau, F., Sterpu, S., Boyer, R., Chan, K., Hannett, L., Leonard, D., Lima, L., Hofbauer, W., Gerin-Lajoie, L., Patterson, S., Undrill, J., Langenbacher, F.: Dynamic Models for Turbine-Governors in Power System Studies
4. Djukanovic, M., Novicevic, M., Dobrijevic, D., Babic, B., Sobajic, D.J., Pao, Y.-H.: Neural-net based coordinated stabilizing control for the exciter and governor loops of low head hydropower plants. IEEE Trans. Energy Convers. **10**(4), 760–767 (1995)
5. Kishor, N., Singh, S.P., Raghuvanshi, A.S., Sharma, P.R.: Fuzzy models for the study of hydro power plant dynamics. In: International Symposium on Evolving Fuzzy Systems, 2006, pp. 235–240
6. Chang, J., Xiao, Z.H. and Wang, S.Q.: Neural network model predict control for the hydroturbine generator set. In: Proceedings of the 2003 International Conference on Machine Learning and Cybernetics (IEEE Cat. No. 03EX693), vol. 1, 2003, pp. 540–543
7. Hemeida, A.M.: Wavelet neural network load frequency controller. Energy Convers. Manag. **46**(9), 1613–1630 (2005)
8. Khodabakhshian, A., Edrisi, M.: A new robust PID load frequency controller. Control Eng. Pract. **16**(9), 1069–1080 (2008)
9. Ying, H.: Fuzzy Control and Modeling: Analytical Foundations and Applications, 1st edn. Wiley, New York
10. Mahmoud, M., Dutton, K., Denman, M.: Design and simulation of a nonlinear fuzzy controller for a hydropower plant. Electr. Power Syst. Res. **73**(2), 87–99 (2005)
11. Yin-Song, W., Guo-Cai, S., Tong-Xiang, H.: The PID-type fuzzy neural network control and it's application in the hydraulic turbine generators. In: 2000 IEEE Power Engineering Society Winter Meeting. Conference Proceedings (Cat. No. 00CH37077), vol. 1, 2000, pp. 338–343
12. Çam, E.: Application of fuzzy logic for load frequency control of hydroelectrical power plants. Energy Convers. Manag. **48**(4), 1281–1288 (2007)
13. Yu, X., Yang, F., Huang, Y., Nan, H.: Fuzzy immune sliding mode control based hydro turbine governor. In: Third International Conference on Natural Computation (ICNC 2007), vol. 1, 2007, pp. 171–176
14. Hiyama, T., Suzuki, N., Karino, H., Lee, K.Y., Andou, H.: Artificial neural network based modeling of governor-turbine system. in: IEEE Power Engineering Society. 1999 Winter Meeting (Cat. No. 99CH36233), vol. 1, 1999, pp. 129–133
15. Babunski, D., Tuneski, A.: Modelling and design of hydraulic turbine–governor system. IFAC Proc. Vol. **36**(7), 263–267 (2003). 3rd IFAC Workshop on Automatic Systems for Building the Infrastructure in Developing Countries 2003, Istanbul, Turkey, 26–28 June 2003. https://doi.org/10.1016/S1474-6670(17)35842-1

16. Li, Y., Peng, C., Yang, Z.: Steam turbine governor modeling and parameters testing for power system simulation. Front. Energy Power Eng. China **3**(2), 198–203 (2009). https://doi.org/10.1007/s11708-009-0004-2
17. Chan, M.L., Dunlop, R.D., Schweppe, F.: Dynamic equivalents for average system frequency behavior following major disturbances. IEEE Trans. Power Appar. Syst. **PAS-91**(4), 1637–1642 (1972)
18. Xie, Y., Zhang, H., Li, C., Sun, H.: Development approach of a programmable and open software package for power system frequency response calculation. Prot. Control Mod. Power Syst. **2**(1), 18 (2017). https://doi.org/10.1186/s41601-017-0045-1
19. Pondini, M., Signorini, A., Colla, V., Barsali, S.: Analysis of a simplified Steam Turbine governor model for power system stability studies. Energy Procedia **158**, 2928–2933 (2019). https://doi.org/10.1016/j.egypro.2019.01.953
20. Umrao, R., Chaturvedi, D.K.: Load frequency control using polar fuzzy controller. In: TENCON 2010—2010 IEEE Region 10 Conference, 2010, pp. 557–562
21. Chamorro, H.R., Sanchez, A.C., Pantoja, A., Zelinka, I., Gonzalez-Longatt, F., Sood, V.K.: A network control system for hydro plants to counteract the non-synchronous generation integration. Int. J. Electr. Power Energy Syst. **105**, 404–419 (2019)
22. Bevrani, H.: Robust Power System Frequency Control, Power Electronics and Power Systems. Springer, New York. https://doi.org/10.1007/978-0-387-84878-5
23. Machowski, J., Lubosny, Z., Bialek, J.W., Bumby, J.R.: Power System Dynamics: Stability and Control, 3rd edn. Wiley, Hoboken, NJ
24. Li, W., Vanfretti, L., Chompoobutrgool, Y.: Development and implementation of hydro turbine and governor models in a free and open source software package. Simul. Model. Pract. Theory **24**, 84–102 (2012)
25. Report, I.C.: Dynamic models for steam and hydro turbines in power system studies. IEEE Trans. Power Appar. Syst. **PAS-92**(6), 1904–1915 (1973)
26. Bezdek, J.C.: Pattern Recognition with Fuzzy Objective Function Algorithms, 1st edn. Plenum Press, New York
27. Bezdek, J.C., Ehrlich, R., Full, W.: FCM: the fuzzy C-means clustering algorithm. Comput. Geosci. **10**(2), 191–203 (1984)
28. Espitia, H.E., Chamorro, H.R., Soriano, J.J.: Fuzzy controller design using concretion based on Boolean relations (CBR). In: 2012 12th UK Workshop on Computational Intelligence (UKCI), 2012, pp. 1–6
29. Barrios, J.A., Cavazos, A., Leduc, L., Ramírez, J.: Fuzzy and fuzzy grey-box modelling for entry temperature prediction in a hot strip mill. Mater. Manuf. Processes **26**(1), 66–77 (2011)
30. K.M. Passino, Yurkovich, S.: Fuzzy Control, 1st edn. Addison-Wesley, Boston
31. Chamorro, H.R., Toro, B.W.: Control de Velocidad de Motores de Inducción y Celdas Resonantes: Aplicando Lógica Difusa, Editorial Académica Española
32. Chamorro, H.R., Riaño, I., Gerndt, R., Zelinka, I., Gonzalez-Longatt, F., Sood, V.K.: Synthetic inertia control based on fuzzy adaptive differential evolution. Int. J. Electr. Power Energy Syst. **105**, 803–813 (2019)
33. Jang, J.S.R., Sun, C.T., Mizutani, E.: Neuro-Fuzzy and Soft Computing: A Computational Approach to Learning and Machine Intelligence. Prentice Hall (1997)
34. Jang, J.R.: ANFIS: adaptive-network-based fuzzy inference system. IEEE Trans. Syst. Man Cybern. **23**(3), 665–685 (1993)
35. Chamorro, H.R., Ordonez, C.A., Peng, J.C., Ghandhari, M.: Non-synchronous generation impact on power systems coherency. IET Gener. Transm. Distrib. **10**(10), 2443–2453 (2016)
36. Mu, Y., Wu, J., Ekanayake, J., Jenkins, N., Jia, H.: Primary frequency response from electric vehicles in the great Britain power system. IEEE Trans. Smart Grid **4**(2), 1142–1150 (2013)
37. Obaid, Z.A., Cipcigan, L., Muhsin, M.T.: Analysis of the great Britain's power system with electric vehicles and storage systems. In: 2015 18th International Conference on Intelligent System Application to Power Systems (ISAP), 2015, pp. 1–6
38. Barrios, J.A., Villanueva, C., Cavazos, A., Colas, R.: Fuzzy C-means rule generation for fuzzy entry temperature prediction in a hot strip mill. J. Iron Steel Res. Int. **23**(2), 116–123 (2016)

Multicriteria Group Decision Making Using a Novel Similarity Measure for Triangular Fuzzy Numbers Based on Their Newly Defined Expected Values and Variances

Palash Dutta and Gourangajit Borah

Abstract This article proposes a novel similarity measure (SM) based on newly defined expected values and variances of triangular fuzzy numbers (TFNs) and hence revealing its applications to group decision making method. The proposed SM approach is based on the idea of defining two expected values and correspondingly their variances for the left and right reference functions of a TFN. The proposed approach is checked for its rationality, by comparing with several existing approaches considering certain sets of TFNs. A multicriteria group decision making method is illustrated, where the values of the criteria are in the form of linguistic values, which can be eventually converted to TFNs. Finally, we consider a practical multicriteria group decision making (MCGDM) scenario, where our proposed method has proved to be practical and in compliance with human intuition. The most desirable alternative is determined based on the similarity measure values between every alternative and an ideal alternative. The applicability of the present approach with existing approaches is validated by obtaining identical preference ordering of alternatives, thus proving its effectiveness.

Keywords Multicriteria group decision making · Triangular fuzzy number · Similarity measure · Vagueness

1 Introduction

In modern decision science, fuzzy set theory is an important aspect [5, 6]. It has found its rigorous applications in the field of economics, engineering technology, military, etc. Multiple criteria decision making under fuzzy environment involves ranking the alternatives based on some predefined multicriteria. Now, the values for these alternative(s) and criteria(s) may be fuzzy sets [31], or intuitionistic fuzzy sets [19, 26],

P. Dutta (✉) · G. Borah
Department of Mathematics, Dibrugarh University, Dibrugarh, Assam 786004, India
e-mail: palashdutta@dibru.ac.in

D. Gupta et al. (eds.), *Soft Computing for Data Analytics, Classification Model, and Control*, Studies in Fuzziness and Soft Computing 413,
https://doi.org/10.1007/978-3-030-92026-5_8

triangular fuzzy numbers [20], interval-valued triangular fuzzy numbers [33], triangular intuitionistic fuzzy numbers [17], trapezoidal fuzzy numbers [1, 27], interval-valued trapezoidal fuzzy numbers [21], intuitionistic trapezoidal fuzzy numbers [28], and the list goes on. Dubois and Prade [8, 9] first introduced the concept of a fuzzy number and the operations involving fuzzy numbers. While, Chen [2] introduced the more generalized trapezoidal fuzzy numbers (GTrFNs). Later on, Dubois and Prade [9] and Heilpern [11] gave the concept of an expected interval of a fuzzy number. Then it was Chen [4], who devised a SM for assessing the subjective mental workload and risk analysis in a fuzzy environment. Consequently, Hsieh and Chen [13] used the "graded mean integration representation distance" to devise a SM between fuzzy numbers. Moreover, Lee [16] proposed SM for trapezoidal fuzzy numbers. Furthermore, Chen and Chen [3] proposed a SM based on the geometric distance concept and the centre of gravity distance for GTrFNs. Also, it was Wei and Chen [25], who combined the concepts of perimeter, height and geometric distances in devising a SM for GTrFNs having applications in fuzzy risk analysis. Moreover, SMs based on distance models like, Hausdorff distance similarity method [12], Hamming distance similarity method [18] and Euclidean distance similarity method [26], were proposed. Several measures were also developed to evaluate similarity between two intuitionistic fuzzy sets [29]. Often in vector spaces, humongous applications of Jaccard, Dice and cosine SMs can be found in the field of citation analysis, automatic classification and information retrieval [7, 14, 22].

TFNs are computationally simple, very intuitive and easy to use. Thus, it finds its applications in FMCDM problems where the criteria values are represented by TFNs. One such work by Fu [10], where he devised a fuzzy optimization method involving ideal and anti-ideal points for solving FMCDM problems considering TFNs. Also, substantial amount of research works could be found in the literature involving TFNs like Wei [23], Wei et al. [24], Zhang et al. [32] and so on.

It is worth mentioning that, Ye [30] applied the Dice SM based on the expected interval for GTrFNs and finally revealed its applicability in MCGDM methods. It motivated us to think of newly defining the expected values and their variances for the left and right reference functions of TFNs separately. So, the main purpose of this article is to devise a novel SM involving the newly defined expected values and variances for the two reference functions of a TFN. Once we have prove that, our proposed approach satisfies the norms of a similarity measure, then a group decision making method is illustrated where the criteria values are represented by TFNs with respect to the linguistic terms.

In order to do so, the rest of the paper is set out as follows. In the next section, we present some basic definitions and arithmetic operational laws of TFNs; three vector similarity measures of TFNs are also discussed. In Sect. 3, we introduce our proposed SM approach. Section 4 presents a comparative analysis of the present approach with the earlier existing approaches, considering seven different sets of TFNs. The results obtained are effective and feasible. Finally, in Sect. 5 we discuss a FMCGDM method and hence validate the proposed SM taking a practical scenario. The paper ends with the concluding remarks in Sect. 6.

2 Preliminaries

We consider the following well-known definitions of a TFN λ.

Definition 1 (*Triangular Fuzzy Number*) [15]

A TFN λ can be defined as (a_1, a_2, a_3) and its membership function is given by,

$$\mu_\lambda(x) = \begin{cases} 0, & x < a_1 \\ \frac{x-a_1}{a_2-a_1}, & a_1 \le x < a_2 \\ \frac{x-a_3}{a_2-a_3}, & a_2 \le x < a_3 \\ 0, & a_3 < x \end{cases} \tag{1}$$

where $0 \le a_1 \le a_2 \le a_3 \le 1$; a_1 and a_3 signify the lower and upper values of the support of λ, whereas a_2 signifies the nodal values. A TFN is pictorially represented in Fig. 1.

Suppose, $\lambda = (a_1, a_2, a_3)$ and $\eta = (b_1, b_2, b_3)$ are two non-zero TFNs and σ be any positive scalar; then some basic arithmetic operations are defined as:

$$\lambda \oplus \eta = (a_1 + b_1, a_2 + b_2, a_3 + b_3) \tag{2}$$

$$\lambda \otimes \eta = (a_1 b_1, a_2 b_2, a_3 b_3) \tag{3}$$

$$\sigma \otimes \lambda = (\sigma a_1, \sigma a_2, \sigma a_3) \tag{4}$$

$$\frac{1}{\lambda} = \left(\frac{1}{a_1}, \frac{1}{a_2}, \frac{1}{a_3}\right) \tag{5}$$

Now, there is a frequent use of some vector similarity measures amongst vector spaces, which are successfully applied to the field of fuzzy assignment problems, image processing, pattern recognition, etc. So, hereby we introduce them as follows.

Fig. 1 TFN, $\lambda = (a_1, a_2, a_3)$

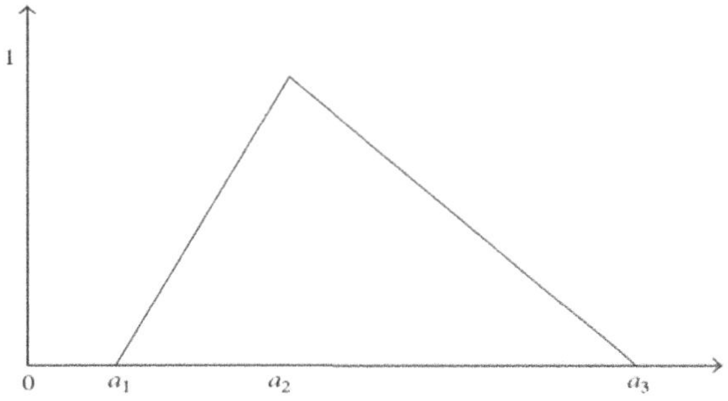

Suppose, $P = (p_1, p_2, \ldots, p_n)$ and $Q = (q_1, q_2, \ldots, q_n)$ be two vectors of length n, and we conclude that all co-ordinates assume positive values, then,

$$\begin{aligned}\text{Jaccard SM, } J(P,Q) &= \frac{PQ}{\|P\|_2^2 + \|Q\|_2^2 - PQ} \\ &= \frac{\sum_{i=1}^n p_i q_i}{\sum_{i=1}^n p_i^2 + \sum_{i=1}^n q_i^2 - \sum_{i=1}^n p_i q_i}\end{aligned} \tag{6}$$

$$\text{Dice SM, } E(P,Q) = \frac{2PQ}{\|P\|_2^2 + \|Q\|_2^2} = \frac{2\sum_{i=1}^n p_i q_i}{\sum_{i=1}^n p_i^2 + \sum_{i=1}^n q_i^2} \tag{7}$$

$$\text{Cosine SM, } C(P,Q) = \frac{PQ}{\|P\|_2 \|Q\|_2} = \frac{\sum_{i=1}^n p_i q_i}{\sqrt{\sum_{i=1}^n p_i^2}\sqrt{\sum_{i=1}^n q_i^2}} \tag{8}$$

Definition 2 (*Expected Interval of GTrFN*) [30]

According to Ye, when $A = (a_1, a_2, a_3, a_4;\ w_A)$ is a GTrFN in the domain of real numbers $\mathbb{R}$, with continuous and strictly monotonic sides, $f_A(x) = \frac{x-a_1}{a_2-a_1} w_A$ and $g_A(x) = \frac{x-a_4}{a_3-a_4} w_A$, then the inverse functions would be respectively,

$$f_A^{-1}(x) = a_1 + (a_2 - a_1)\frac{y}{w_A}, \quad y \in [0, w_A] \tag{9}$$

$$g_A^{-1}(x) = a_4 + (a_3 - a_4)\frac{y}{w_A}, \quad y \in [0, w_A] \tag{10}$$

Hence, the expected interval of a GTrFN $A = (a_1, a_2, a_3, a_4;\ w_A)$ is the crisp interval $EI(A)$, where

$$\begin{aligned}EI(A) &= [E_L(A), E_U(A)] \\ &= \left[\int_0^{w_A} f_A^{-1}(x)dy, \int_0^{w_A} g_A^{-1}(x)dy\right]\end{aligned}$$

Now, very clearly,

$$E_L(A) = \int_0^{w_A} \left[a_1 + (a_2 - a_1)\frac{y}{w_A}\right] dy$$

$$\Rightarrow E_L(A) = a_1 w_A + \frac{a_2 - a_1}{2} w_A = \frac{w_A(a_1 + a_2)}{2}$$

and,

$$E_U(A) = \int_0^{w_A} \left[a_4 + (a_3 - a_4)\frac{y}{w_A}\right]dy$$

$$\Rightarrow E_U(A) = a_4 w_A + \frac{a_3 - a_4}{2} w_A = \frac{w_A(a_3 + a_4)}{2}$$

So, the expected interval is,

$$EI(A) = \left[\frac{w_A}{2}(a_1 + a_2), \frac{w_A}{2}(a_3 + a_4)\right] \quad (11)$$

Then, the expected value is defined to be the centre-point of the expected interval of A:

$$E(A) = \frac{E_L(A) + E_U(A)}{2} = \frac{w_A(a_1 + a_2 + a_3 + a_4)}{4} \quad (12)$$

Definition 3

Motivated by Ye [30], we have made an attempt to define separate expected values and their variances for the left and right reference functions/sides of a TFN. That is, let $\lambda = (a_1, a_2, a_3)$ be a TFN, having the membership function as,

$$\mu_\lambda(x) = \begin{cases} p(x), & a_1 \le x < a_2 \\ q(x), & a_2 \le x < a_3 \\ 0, & otherwise \end{cases} \quad (13)$$

Here, $p(x) = \frac{x - a_1}{a_2 - a_1}$ is known as the Left Reference Function (L.R.F.) of the TFN λ, and similarly, $q(x) = \frac{x - a_3}{a_2 - a_3}$ is known as the Right Reference Function (R.R.F.) of the TFN λ.

Expected value and Variance on the L.R.F. of λ:

We define,

Expected value for L.R.F.,

$$E_L(\lambda) = \int_0^1 [a_1 + (a_2 - a_1)x]dx \quad (14)$$

Variance for L.R.F.,

$$V_L(\lambda) = E_L(\lambda^2) - [E_L(\lambda)]^2 \quad (15)$$

It is to be noted,

$$\lambda^2 = \left(a_1^2, a_2^2, a_3^2\right)$$

From (14),

$$\begin{aligned} E_L(\lambda) &= \left[a_1 x + (a_2 - a_1)\frac{x^2}{2}\right]_0^1 \\ &= a_1 + (a_2 - a_1)\frac{1}{2} \\ &= \frac{2a_1 + a_2 - a_1}{2} \\ \Rightarrow E_L(\lambda) &= \frac{a_1 + a_2}{2} \end{aligned}$$

So,

$$E_L(\lambda^2) = \frac{a_1^2 + a_2^2}{2}$$

Therefore,

$$\begin{aligned} V_L(\lambda) &= E_L\left(\lambda^2\right) - [E_L(\lambda)]^2 \\ &= \frac{a_1^2 + a_2^2}{2} - \left[\frac{a_1 + a_2}{2}\right]^2 \\ &= \frac{a_1^2 + a_2^2}{2} - \frac{1}{4}(a_1 + a_2)^2 \\ &= \frac{2a_1^2 + 2a_2^2 - a_1^2 - a_2^2 - 2a_1 a_2}{4} \\ &= \frac{a_1^2 + a_2^2 - 2a_1 a_2}{4} \\ \Rightarrow V_L(\lambda) &= \frac{(a_1 - a_2)^2}{4} \end{aligned} \quad (16)$$

Expected value and Variance on the R.R.F. of λ:

By following the similar calculations done for the L.R.F. of λ, we can obtain the values for R.R.F. as,

Expected value for R.R.F.,

$$E_R(\lambda) = \int_0^1 [a_3 + (a_2 - a_3)x]dx = \frac{a_2 + a_3}{2} \quad (17)$$

Variance for R.R.F.,

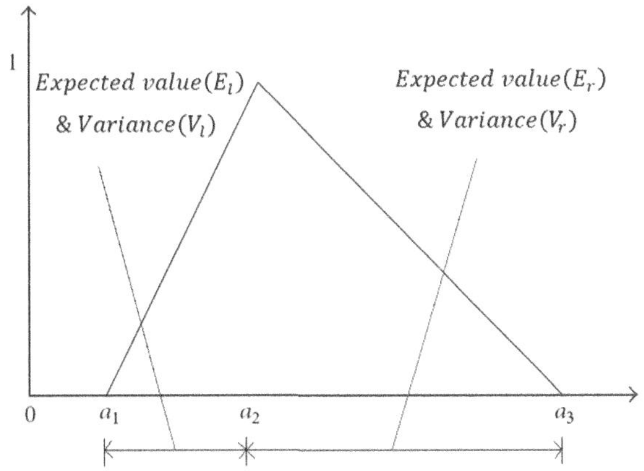

Fig. 2 Figure depicting $E_L(\lambda)$, $E_R(\lambda)$, $V_L(\lambda)$, $V_R(\lambda)$ for TFN $\lambda = (a_1, a_2, a_3)$

$$V_R(\lambda) = E_R(\lambda^2) - [E_R(\lambda)]^2 = \frac{(a_2 - a_3)^2}{4} \tag{18}$$

A diagram will help us have a clear view in this regard, as shown in Fig. 2.

3 Proposed SM Approach

Let us consider two non-zero TFNs, $\lambda = (a_1, a_2, a_3)$ and $\eta = (b_1, b_2, b_3)$, where $0 \le a_1 \le a_2 \le a_3 \le 1$; $0 \le b_1 \le b_2 \le b_3 \le 1$. Then, the similarity measure between λ and η is proposed as,

$$S^N(\lambda, \eta) = \frac{2\left[\sum_{i=1}^{3} a_i b_i + E_L(\lambda)E_L(\eta) + E_R(\lambda)E_R(\eta) + V_L(\lambda)V_L(\eta) + V_R(\lambda)V_R(\eta)\right]}{\sum_{i=1}^{3}\left(a_i^2 + b_i^2\right) + E_L(\lambda)^2 + E_L(\eta)^2 + E_R(\lambda)^2 + E_R(\eta)^2 + V_L(\lambda)^2 + V_L(\eta)^2 + V_R(\lambda)^2 + V_R(\eta)^2} \tag{19}$$

where $a_i, b_i (i = 1, 2, 3)$, $E_L(\lambda)$, $E_R(\lambda)$- expected values of the reference functions of λ, $E_L(\eta)$, $E_R(\eta)$—expected values of the reference functions of η, $V_L(\lambda)$, $V_R(\lambda)$—variance values of the reference functions of λ, $V_L(\eta)$, $V_R(\eta)$—variance values of the reference functions of η.

Theorem 1 To prove, $S^N(\lambda, \eta)$ as a similarity measure, it must satisfy the following properties:

P-1: $0 \le S^N(\lambda, \eta) \le 1$.
P-2: $S^N(\lambda, \eta) = S^N(\eta, \lambda)$.
P-3: $S^N(\lambda, \eta) = 1 \Leftrightarrow \lambda = \eta$.

Proof (**P-1**) Clearly by definition, $S^N(\lambda, \eta) \ge 0$.

Now,

$$2\left[\sum_{i=1}^{3} a_i b_i + E_L(\lambda)E_L(\eta) + E_R(\lambda)E_R(\eta) + V_L(\lambda)V_L(\eta) + V_R(\lambda)V_R(\eta)\right]$$

$$\leq \sum_{i=1}^{3}\left(a_i^2 + b_i^2\right) + E_L(\lambda)^2 + E_L(\eta)^2 + E_R(\lambda)^2 + E_R(\eta)^2 + V_L(\lambda)^2 + V_L(\eta)^2$$

$$+ V_R(\lambda)^2 + V_R(\eta)^2$$

$$\Rightarrow 0 \leq S^N(\lambda, \eta) \leq 1.$$

(P-2) Here,

$$S^N(\lambda,\eta) = \frac{2\left[\sum_{i=1}^{3} a_i b_i + E_L(\lambda)E_L(\eta) + E_R(\lambda)E_R(\eta) + V_L(\lambda)V_L(\eta) + V_R(\lambda)V_R(\eta)\right]}{\sum_{i=1}^{3}\left(a_i^2 + b_i^2\right) + E_L(\lambda)^2 + E_L(\eta)^2 + E_R(\lambda)^2 + E_R(\eta)^2 + V_L(\lambda)^2 + V_L(\eta)^2 + V_R(\lambda)^2 + V_R(\eta)^2}$$
$$= \frac{2\left[\sum_{i=1}^{3} b_i a_i + E_L(\eta)E_L(\lambda) + E_R(\eta)E_R(\lambda) + V_L(\eta)V_L(\lambda) + V_R(\eta)V_R(\lambda)\right]}{\sum_{i=1}^{3}\left(b_i^2 + a_i^2\right) + E_L(\eta)^2 + E_L(\lambda)^2 + E_R(\eta)^2 + E_R(\lambda)^2 + V_L(\eta)^2 + V_L(\lambda)^2 + V_R(\eta)^2 + V_R(\lambda)^2}$$
$$= S^N(\eta, \lambda)$$

(P-3) For $\lambda = \eta$, we have $a_i = b_i$, $E_L(\lambda) = E_L(\eta)$, $E_R(\lambda) = E_R(\eta)$, $V_L(\lambda) = V_L(\eta)$, $V_R(\lambda) = V_R(\eta)$, then clearly, $S^N(\lambda, \eta) = 1$.

Conversely, let $S^N(\lambda, \eta) = 1$, then

$$2\left[\sum_{i=1}^{3} a_i b_i + E_L(\lambda)E_L(\eta) + E_R(\lambda)E_R(\eta) + V_L(\lambda)V_L(\eta) + V_R(\lambda)V_R(\eta)\right]$$

$$= \sum_{i=1}^{3}\left(a_i^2 + b_i^2\right) + E_L(\lambda)^2 + E_L(\eta)^2 + E_R(\lambda)^2 + E_R(\eta)^2 + V_L(\lambda)^2$$

$$+ V_L(\eta)^2 + V_R(\lambda)^2 + V_R(\eta)^2$$

$$\Rightarrow \sum_{i=1}^{3}(a_i - b_i)^2 + (E_L(\lambda) - E_L(\eta))^2 + (E_R(\lambda) - E_R(\eta))^2 + (V_L(\lambda) - V_L(\eta))^2$$

$$+ (V_R(\lambda) - V_R(\eta))^2 = 0, \text{ for } i = 1, 2, 3, \text{ otherwise } S^N(\lambda, \eta) \neq 1, \text{ which is a contradiction.}$$

$$\Rightarrow \lambda = \eta$$

Hence, $S^N(\lambda, \eta)$ is a similarity measure.

4 Comparative Analysis of the Similarity Measures

Here, in this section we consider seven sets of TFNs discussed in Ye [30] as shown in Table 1.

Table 1 Seven sets of TFNs

Sets	λ	η
Set 1	(0.1, 0.2, 0.3)	(0.1, 0.2, 0.3)
Set 2	(0.3, 0.3, 0.3)	(0.3, 0.3, 0.3)
Set 3	(0.2, 0.2, 0.2)	(0.3, 0.3, 0.3)
Set 4	(0.1, 0.2, 0.3)	(0.3, 0.3, 0.3)
Set 5	(0.1, 0.2, 0.3)	(0.2, 0.3, 0.4)
Set 6	(0.1, 0.4, 0.7)	(0.3, 0.4, 0.5)
Set 7	(0.4, 0.4, 0.8)	(0.3, 0.4, 0.5)

Table 2 Comparison of the evaluated results of the proposed method with the already existing methods

Sets	Chen's method [4]	Lee's method [16]	Hsieh and Chen's method [13]	Chen and Chen's method [3]	Wei and Chen [25]	Ye [30]	Proposed approach
Set 1	1	1	1	1	1	1	1
Set 2	1	#	1	1	1	1	1
Set 3	0.9	0	0.9091	0.81	0.9	0.9231	0.9230
Set 4	*0.9*	0.5	*0.9091*	0.54	0.8411	0.9057	0.8889
Set 5	*0.9*	0.6667	*0.9091*	0.81	0.9	0.9259	0.9286
Set 6	0.9	0.8333	*1*	*0.9*	0.7833	0.9710	0.9455
Set 7	0.9	0.8	0.9375	0.8325	0.8289	0.9704	0.9452

"#"—denotes that the degree of similarity cannot be determined by the approach; values in italics are incorrect

We then compare the evaluated results of the proposed method with the outcomes of the existing similarity measures [3, 4, 13, 16, 25, 30] as shown in Table 2.

Very evident from Table 2, we can observe that the similarity measures by Chen's method, Lee's method, Hsieh and Chen's method and Chen and Chen's method had the drawbacks by producing incorrect or absurd results in some cases. However, Wei and Chen's method, Ye's method and the proposed method possess the satisfying property for identification of similarity and can overcome the drawbacks of the other similarity measures.

5 FMCGDM Method with the Proposed SM Approach

Here, in this section, we discuss multiple criteria group decision making problems using TFNs.

Table 3 Linguistic values of TFNs

Linguistic term	Linguistic values of TFNs
Absolutely poor (AP)	(0.0, 0.0, 0.1)
Very poor (VP)	(0.0, 0.1, 0.2)
Poor (P)	(0.1, 0.2, 0.3)
Medium poor (MP)	(0.3, 0.4, 0.5)
Medium (M)	(0.4, 0.5, 0.6)
Medium good (MG)	(0.5, 0.6, 0.7)
Good (G)	(0.7, 0.8, 0.9)
Very good (VG)	(0.8, 0.9, 1.0)
Absolutely good (AG)	(0.9, 1.0, 1.0)

Let $A = \{A_1, A_2, \ldots, A_p\}$ be the set of alternative(s) and let $C = \{C_1, C_2, \ldots, C_q\}$ be the set of criteria(s), respectively. Let us assume k experts are called upon to make the judgement and it is to be noted that, we are considering criteria(s) to be of equal weightage. The experts are expected to give the linguistic value of TFNs (see Table 3).

Now, the vectors of the alternative A_i, given by the k experts are given by following TFNs:

$$v_i^s = \{\langle C_1, (a_{i11}^s, a_{i12}^s, a_{i13}^s)\rangle, \langle C_2, (a_{i21}^s, a_{i22}^s, a_{i23}^s)\rangle, \ldots, \langle C_q, (a_{iq1}^s, a_{iq2}^s, a_{iq3}^s)\rangle\} \tag{20}$$

where $a_{ijm} \in \mathbb{R}(i = 1, 2, \ldots, p;\ j = 1, 2, \ldots, q;\ m = 1, 2, 3;\ s = 1, 2, \ldots, k)$. Thus, the procedure for selecting the best alternative or preference ordering of alternatives is presented as follows.

Step-1: Now, we construct the group preference vector based on the values given by the experts,

$$V_i = \left\{\left\langle C_1, \left(\frac{\sum_{s=1}^k a_{i11}^s}{k}, \frac{\sum_{s=1}^k a_{i12}^s}{k}, \frac{\sum_{s=1}^k a_{i13}^s}{k}\right)\right\rangle, \left\langle C_2, \left(\frac{\sum_{s=1}^k a_{i21}^s}{k}, \frac{\sum_{s=1}^k a_{i22}^s}{k}, \frac{\sum_{s=1}^k a_{i23}^s}{k}\right)\right\rangle, \ldots, \left\langle C_q, \left(\frac{\sum_{s=1}^k a_{iq1}^s}{k}, \frac{\sum_{s=1}^k a_{iq2}^s}{k}, \frac{\sum_{s=1}^k a_{iq3}^s}{k}\right)\right\rangle\right\} \tag{21}$$

Thus, we can elicit these values in the form of a decision matrix.

Step-2: Although realizing an ideal alternative is hard in the real world. But it is necessary as it provides a useful theoretical construct for evaluating alternatives [30]. The decision maker sets the ideal alternative as per the criteria value(s).

Step-3: Once the ideal alternative is set by the decision maker, we can define the vector similarity measures between an alternative, A_i and the ideal alternative, A^*, represented by TFNs as follows,

$$S^J(A_i, A^*) = \frac{1}{q}\sum_{j=1}^{q} \frac{\sum_{m=1}^{3} a_{ijm}a_{*jm}}{\sum_{m=1}^{3} a_{ijm}^2 + \sum_{m=1}^{3} a_{*jm}^2 - \sum_{m=1}^{3} a_{ijm}a_{*jm}} \tag{22}$$

$$S^E(A_i, A^*) = \frac{1}{q}\sum_{j=1}^{q} \frac{2\sum_{m=1}^{3} a_{ijm}a_{*jm}}{\sum_{m=1}^{3} a_{ijm}^2 + \sum_{m=1}^{3} a_{*jm}^2} \tag{23}$$

$$S^C(A_i, A^*) = \frac{1}{q}\sum_{j=1}^{q} \frac{\sum_{m=1}^{3} a_{ijm}a_{*jm}}{\sqrt{\sum_{m=1}^{3} a_{ijm}^2}\sqrt{\sum_{m=1}^{3} a_{*jm}^2}} \tag{24}$$

Since, we are interested to check the effectiveness of our proposed method with [30] approach too, so in addition to SMs mentioned above, we mention about Ye's approach which is,

$$S^Y(A_i, A^*) = \frac{1}{q}\sum_{j=1}^{q} \frac{2\left[\left(\frac{a_{ij1}+a_{ij2}}{2}\right)\left(\frac{a_{*j1}+a_{*j2}}{2}\right) + \left(\frac{a_{ij2}+a_{ij3}}{2}\right)\left(\frac{a_{*j2}+a_{*j3}}{2}\right)\right]}{\left[\left(\frac{a_{ij1}+a_{ij2}}{2}\right)^2 + \left(\frac{a_{ij2}+a_{ij3}}{2}\right)^2 + \left(\frac{a_{*j1}+a_{*j2}}{2}\right)^2 + \left(\frac{a_{*j2}+a_{*j3}}{2}\right)^2\right]} \tag{25}$$

and finally, our proposed method which takes the form,

$$S^N(A_i, A^*) = \frac{1}{q}\sum_{j=1}^{q} \frac{2\left[\sum_{m=1}^{3} a_{ijm}a_{*jm} + E_L(A_i)E_L(A^*) + E_R(A_i)E_R(A^*) + V_L(A_i)V_L(A^*) + V_R(A_i)V_R(A^*)\right]}{\sum_{m=1}^{3}\left(a_{ijm}^2 + a_{*jm}^2\right) + E_L(A_i)^2 + E_L(A^*)^2 + E_R(A_i)^2 + E_R(A^*)^2 + V_L(A_i)^2 + V_L(A^*)^2 + V_R(A_i)^2 + V_R(A^*)^2} \tag{26}$$

Step-4: We rank the alternatives and then select the best one in accordance with the values of evaluated SMs.

5.1 *Ilustrated Example*

In this section, we illustrate the application of our proposed method in a practical scenario of personnel selection (adapted from Zhang et al. [32]).

Suppose a competent is desired by the human resources department of a company. Four candidates (i.e., alternatives) A_1, A_2, A_3 and A_4 are left for further selection, after the process of initial screening. Foe selecting the most desirable candidate, the decision maker invites 3(three) experts who evaluate the alternatives based on the following four criteria: (1) emotional steadiness (C_1); (2) education experience (C_2); (3) work experience (C_3); (4) personality and self-confidence (C_4). We assume that,

Table 4 Preference values of alternatives given by three experts in the form of TFNs

	K	C_1	C_2	C_3	C_4
A_1	1	(0.0, 0.0, 0.1)	(0.4, 0.5, 0.6)	(0.5, 0.6, 0.7)	(0.5, 0.6, 0.7)
	2	(0.3, 0.4, 0.5)	(0.1, 0.2, 0.3)	(0.0, 0.1, 0.2)	(0.7, 0.8, 0.9)
	3	(0.5, 0.6, 0.7)	(0.7, 0.8, 0.9)	(0.2, 0.3, 0.4)	(0.5, 0.6, 0.7)
A_2	1	(0.3, 0.4, 0.5)	(0.1, 0.2, 0.3)	(0.7, 0.8, 0.9)	(0.5, 0.6, 0.7)
	2	(0.1, 0.2, 0.3)	(0.0, 0.0, 0.1)	(0.5, 0.6, 0.7)	(0.0, 0.1, 0.2)
	3	(0.4, 0.5, 0.6)	(0.4, 0.5, 0.6)	(0.3, 0.4, 0.5)	(0.7, 0.8, 0.9)
A_3	1	(0.4, 0.5, 0.6)	(0.8, 0.9, 1.0)	(0.5, 0.6, 0.7)	(0.0, 0.1, 0.2)
	2	(0.5, 0.6, 0.7)	(0.7, 0.8, 0.9)	(0.3, 0.4, 0.5)	(0.0, 0.0, 0.1)
	3	(0.5, 0.6, 0.7)	(0.3, 0.4, 0.5)	(0.4, 0.5, 0.6)	(0.8, 0.9, 1.0)
A_4	1	(0.5, 0.6, 0.7)	(0.5, 0.6, 0.7)	(0.4, 0.5, 0.6)	(0.8, 0.9, 1.0)
	2	(0.1, 0.2, 0.3)	(0.0, 0.1, 0.2)	(0.5, 0.6, 0.7)	(0.1, 0.2, 0.3)
	3	(0.7, 0.8, 0.9)	(0.4, 0.5, 0.6)	(0.1, 0.2, 0.3)	(0.3, 0.4, 0.5)

the three experts provide his/her preference value on alternatives with respect to a certain criterion by using linguistic variables as listed in Table 3.

Step-1: Three experts give his/her preference information with linguistic terms, which are converted to TFNs as shown in Table 4.
Step-2: The preference vector as given by the experts for each alternative $A_i (i = 1, 2, 3, 4)$ is depicted in Table 5, in the form of a decision matrix.
Step-3: The ideal alternative is set by the decision maker to be

$$A^* = \{\langle C_1, (0.4, 0.5, 0.6)\rangle, \langle C_2, (0.4, 0.5, 0.5)\rangle, \langle C_3, (0.3, 0.4, 0.6)\rangle, \langle C_4, (0.4, 0.5, 0.6)\rangle\}$$

Step-4: From Table 6, we can observe that the proposed method produces identical decision results. Hence, the decision results of different similarity measures, suggests the rationality and feasibility of the proposed SM approach.

Table 5 Decision matrix for group preference vectors

	C_1	C_2	C_3	C_4
A_1	(0.27, 0.34, 0.44)	(0.4, 0.5, 0.6)	(0.24, 0.34, 0.44)	(0.57, 0.67, 0.77)
A_2	(0.27, 0.37, 0.47)	(0.17, 0.24, 0.34)	(0.5, 0.6, 0.7)	(0.4, 0.5, 0.6)
A_3	(0.47, 0.57, 0.67)	(0.6, 0.7, 0.8)	(0.3, 0.5, 0.6)	(0.27, 0.34, 0.44)
A_4	(0.44, 0.54, 0.64)	(0.3, 0.4, 0.5)	(0.34, 0.44, 0.54)	(0.4, 0.5, 0.6)

Table 6 Decision results of various similarity measures for TFNs

	S^J	S^E	S^C	S^Y	S^N
A_1	0.9329	0.9650	0.9981	0.9670	0.9658
A_2	0.8823	0.9339	0.9919	0.9300	0.9323
A_3	0.9275	0.9615	0.9979	0.9628	0.9621
A_4	0.9873	0.9936	0.9991	0.9946	0.9940
Ranking	$A_4 > A_1 > A_3 > A_2$	$A_4 > A_1 > A_3 > A_2$	$A_4 > A_1 > A_3 > A_2$	$A_4 > A_1 > A_3 > A_2$	$A_4 > A_1 > A_3 > A_2$
Best	A_4	A_4	A_4	A_4	A_4

6 Concluding Remarks

Although the similarity measure proves to be an effective method for solving FMCDM problems, yet it has rarely any applications in triangular FMCDM problems. This article proposed a novel SM between TFNs, by newly defining separate expected value(s) and variance(s) for TFNs. Multicriteria group decision making method is then illustrated, where criteria values are given with the linguistic values of TFNs with respect to the linguistic terms/variables. The simplicity and effectiveness of the proposed method is reflected in considering a practical scenario of personnel selection [32]. The decision results obtained by different methods implicates that our proposed SM approach in this paper, is applicable and feasible.

In the future direction, we will try to extend the proposed measure to triangular intuitionistic fuzzy numbers, triangular neutrosophic fuzzy numbers, which are believed to tackle problems in the field of complex decision making, pattern recognition, medical diagnosis, etc., with a greater level of accuracy.

Acknowledgements **Author Contribution** G.B. wrote the manuscript while P.D. conceived the study and revised the manuscript. Both the author(s) have read and approved the manuscript.

Disclosure statement No potential conflict of interest was reported by the author(s).

Funding This research did not receive any specific grant from funding agencies in the public, commercial, or not-for-profit sectors.

References

1. Abbasbandy, S., Hajjari, T.: A new approach for ranking of trapezoidal fuzzy numbers. Comput. Math. Appl. **57**(3), 413–419 (2009)
2. Chen, S.H.: Fuzzy numbers with maximizing set and minimizing set. Fuzzy Sets Syst. **17**(2), 113–129 (1985)
3. Chen, S.J., Chen, S.M.: A new method to measure the similarity between fuzzy numbers. In: Proceedings of the 10th IEEE International Conference on Fuzzy Systems, 2–5 Dec, Melbourne, Australia, pp. 1123–1126 (2001)
4. Chen, S.M.: New methods for subjective mental workload assessment and fuzzy risk analysis. Cybern. Syst. **27**(5), 449–472 (1996)
5. Chen, T.Y.: Multi-criteria decision-making method with leniency reduction based on interval-valued fuzzy sets. J. Chinese Inst. Ind. Eng. **28**(1), 1–19 (2011a)
6. Chen, T.Y.: Experimental analysis of score functions in multi-criteria decision-making with intuitionistic fuzzy sets. J. Chinese Inst. Ind. Eng. **28**(3), 206–225 (2011b)
7. Dice, L.R.: Measures of the amount of ecologic association between species. Ecology **26**(3), 297–302 (1945)
8. Dubois, D., Prade, H.: Ranking fuzzy number in the setting of possibility theory. Inf. Sci. **30**, 183–224 (1983)
9. Dubois, D., Prade, H.: The mean value of a fuzzy number. Fuzzy Sets Syst. **24**, 279–300 (1987)
10. Fu, G.T.: A fuzzy optimization method for multicriteria decision making: an application to reservoir flood control operation. Expert Syst. Appl. **34**(1), 145–149 (2008)
11. Heilpern, S.: The expected value of a fuzzy number. Fuzzy Sets Syst. **47**, 81–86 (1992)

12. Hung, W.L., Yang, M.S.: Similarity measures of intuitionistic fuzzy sets based on Hausdorff distance. Pattern Recogn. Lett. **25**(14), 1603–1611 (2004)
13. Hsieh, C.H., Chen, S.H.: Similarity of generalized fuzzy numbers with graded mean integration representation. In: Proceedings of the Eighth International Fuzzy Systems Association World Congress, 17–20 Aug, Taipei, Taiwan, Republic of China, 2, pp. 551–555 (1999)
14. Jaccard, P.: Distribution de la flore alpine dans le basin des Dranses et dans quelques regions voisines. Bull Soc Vaudoise Sci Nat **37**, 241–272 (1901)
15. Laarhoven, P.J.M., Pedrycz, W.: A fuzzy extension of Saaty's priority theory. Fuzzy Sets Syst. **11**(3), 229–241 (1983)
16. Lee, H.S.: An optimal aggregation method for fuzzy opinions of group decision. In: Proceedings of the 1999 IEEE International Conference on Systems and Man, and Cybernetics, 12–15 Oct, Tokyo, Japan, 3, pp. 314–319 (1999)
17. Li, D.F.: A ratio ranking method of triangular intuitionistic fuzzy numbers and its application to MADM problems. Comput. Math. Appl. **60**(6), 1557–1570 (2010)
18. Liang, Z., Shi, P.: Similarity measures on intuitionistic fuzzy sets. Pattern Recogn. Lett. **24**, 2678–2693 (2003)
19. Liu, H.W.: New similarity measures between intuitionistic fuzzy sets and between elements. Math. Comput. Model. **42**(1–2), 61–70 (2005)
20. Liu, P.D., Wang, T.J.: Method for multiple attribute decision making with triangular fuzzy number and partial attribute weight information. J. Inf. Comput. Sci. **4**(3), 1017–1022 (2007)
21. Liu, P.D.: A weighted aggregation operators multi-attribute group decision-making method based on interval-valued trapezoidal fuzzy numbers. Expert Syst. Appl. **38**(1), 1053–1060 (2011)
22. Salton, G., McGill, M.J.: Introduction to Modern Information Retrieval. McGraw-Hill, New York, NY (1987)
23. Wei, G.W.: FIOWHM operator and its application to multiple attribute group decision making. Expert Syst. Appl. **38**(4), 2984–2989 (2011)
24. Wei, G.W., Zhao, X.F., Lin, R., Wang, H.J.: Generalized triangular fuzzy correlated averaging operator and their application to multiple attribute decision making. Appl. Math. Model. **36**(7), 2975–2982 (2012)
25. Wei, S.H., Chen, S.M.: A new approach for fuzzy risk analysis based on similarity measures of generalized fuzzy numbers. Expert Syst. Appl. **36**, 589–598 (2009)
26. Xu, Z.S.: Some similarity measures of intuitionistic fuzzy sets and their applications to multiple attribute decision making. Fuzzy Optim. Decis. Making **6**(2), 109–121 (2007)
27. Xu, Z.Y., Shang, S.C., Qian, W.B., Shu, W.H.: A method for fuzzy risk analysis based on the new similarity of trapezoidal fuzzy numbers. Expert Syst. Appl. **37**(3), 1920–1927 (2011)
28. Ye, J.: Expected value method for intuitionistic trapezoidal fuzzy multicriteria decision-making problems. Expert Syst. Appl. **38**(9), 11730–11734 (2011a)
29. Ye, J.: Cosine similarity measures for intuitionistic fuzzy sets and their applications. Math. Comput. Model. **53**(1–2), 91–97 (2011b)
30. Ye, J.: The dice similarity measure between generalized trapezoidal fuzzy numbers based on the expected interval and its multicriteria group decision-making method. J. Chinese Inst. Ind. Eng. **29**(6), 375–382 (2012)
31. Zadeh, L.A.: Fuzzy sets. Inf. Comput. **8**, 338–353 (1965)
32. Zhang, L., Xu, X., Tao, L.: Some similarity measures for triangular fuzzy number and their applications in multi criteria group decision-making. J. Appl. Math. **3** (2013)
33. Zhang, S.F., Liu, S.Y., Zhai, R.H.: An extended GRA method for MCDM with interval-valued triangular fuzzy assessments and unknown weights. Comput. Ind. Eng. **81**, 1336–1341 (2011)

Bangla Printed Character Generation from Handwritten Character Using GAN

Md. Min-ha-zul Abedin, Tapotosh Ghosh, Tazqia Mehrub, and Mohammad Abu Yousuf

Abstract Bangla language contains a vast character set containing more than twelve thousand compound characters. These characters are the key to make words. A lot of words are now on the verge of extinction because Unicode contains only three hundred Bangla compound characters. Another option to save these words from extinction is improving the optical character recognition system. But, Optical Character Recognition (OCR) needs a lot of samples from the diversely distributed dataset for training. Making diverse datasets for OCR training is a challenging and tedious task to perform. In this paper, Generative Adversarial Net (GAN) based architecture capable of generating printed-character from handwritten Bangla character is proposed. GAN model works on the principle of an image to image translation based on image mapping to output. The architecture not only learns a mapping between handwritten-character image and printed-character image but also learn loss function, which is vital to generate printed-character from different probability distribution like Bangla script. In most of the cases, there is hardly any differences found between generated character using GAN and printed-character. The probability distribution of generated printed-character has diversity because of GAN loss. Generated printed-character can be used in training OCR because of variations in the data distribution. Hardcoded loss function or mapping function is no longer needed for handwritten-character to printed-character generation.

Md. M. Abedin (✉)
Department of Information and Communication Engineering, Bangladesh Army University of Engineering and Technology, Qadirabad, Natore, Bangladesh
e-mail: 16511024@student.bup.edu.bd

T. Ghosh · T. Mehrub
Department of Information and Communication Technology, Bangladesh University of Professionals, Dhaka, Bangladesh
e-mail: 16511038@student.bup.edu.bd

T. Mehrub
e-mail: 16511041@student.bup.edu.bd

M. A. Yousuf
Institute of Information Technology, Jahangirnagar University, Savar, Dhaka, Bangladesh
e-mail: yousuf@juniv.edu

D. Gupta et al. (eds.), *Soft Computing for Data Analytics, Classification Model, and Control*, Studies in Fuzziness and Soft Computing 413,
https://doi.org/10.1007/978-3-030-92026-5_9

Keywords GAN · OCR · Bangla compound handwritten characters · Dataset

1 Introduction

Bangla language is the primary language for more than 24 million people residing in the West Bengal region of India and Bangladesh. To represent the Bangla script in modern-day computing, Unicode is used. There are several limitations in encoding Bangla script to Unicode. The number of Unicode supported characters is limited, and the allocation of Unicode for a specific letter of any language takes quite a long time. Bangla script has numerous (more than twelve thousand) compound characters. It is not possible to allocate a Unicode for all compound Bangla characters.

As a consequence of this, many Bangla words are in danger of disappearing, which prevents us from expressing our fullest. A word in any language involves an emotion that was felt by the inhabitants of a specific region. Losing a word is losing an emotion that may have caused our ancestors to be happy or sad, angry, or beloved.

OCR can be a way of saving these characters, but creating a robust OCR is very hard as the probability distribution of character is diverse. OCR based on Artificial Neural Net showed great potential, but a lot of samples are needed from a various probability distribution, which is a significant problem. Making a vast dataset and maintaining diversity is a challenging task.

In this work, we have proposed a GAN based architecture that is capable of generating printed character image from it's corresponding handwritten character image. We have built two models (Generator model and Discriminative model) to perform this work. We have also created a printed character dataset that contains 231 printed characters in SUTONI MJ font. We were successful in generating printed character images from the handwritten character images where the generated images and the target images looked very similar.

2 Literature Review

Image generation is an active arena of research in the field of deep learning. GAN was introduced by Ian Goodfellow et al., which changed the image generation technique and altered the expectations from computer-generated images [1]. GAN is the most successful framework for generating artificial data (image, audio, video, etc.). GAN is a framework that can estimate the generative model through the adversarial process. In this framework, two models (Generative model G and Discriminative model D) are simultaneously trained. G model tries to capture the probability distribution of the provided data, on the other hand, the D model works to evaluate the probability of a sample from the artificially generated data generated by the G model or target data. In the training of G, the goal is the maximization of the probability of D model being false. D model learns to distinguish whether the data is from the G model or target data. The Generator uses the feedback produced by the Discriminator to generate an image or content that is unable to detect as fake image by the Discriminator. The race between Generator and Discriminator continues until they attain equilibrium in which the Discriminator can not detect the generated image as fake.

Variational Autoencoders (VAEs) and other deep learning/AI models are famous for image generation. These models mostly depend on objective likelihood function for their optimization and generate blurred outputs due to input noise and imperfect reconstruction [2]. On the other hand, GAN can produce a high-resolution image, similar to the input, even without a well-defined objective function. GAN mostly concerns training and variation in input data distribution since GAN faces difficulty in operating in low variation input data distribution.

There are many GAN based techniques for image generation. DCGAN [3], PGGAN [4] concerns in generating realistic images from noise without having any control over the generated output image. GAN is also used in Text to Image Synthesis [5], AttnGAN [6] is used for Caption to image generation. Moreover, Sketch to Image generation technique is also available using GAN [7]. Pix2Pix [8], Pix2PixHD [9], CycleGAN [10] are used to generate images from Edge, and these image translation techniques generate an impressive result which was never achieved previously. Scribbler [11], Auto-painter [12], Colorization [13] are more advanced from previous GAN based work for Edge to Image generation and incorporates color to generate an image for them. This technique has more control over the output generated image for any given input image.

GAN based architecture generated images are mainly used in training other neural nets as CNN requires a massive number of variations in the training dataset. Han et al. proposed a model for generating synthetic MR image [14]. Jeng et al. [15] has proposed a Chinese character multi-font Generator based on GAN. Besides this, the calligraphy technique has been addressed in that work. GAN based graph matching model was proposed by Li et al. [16]. Zinnia and Shopon [17] worked on the GAN model to generate real-looking Bangla Handwritten Character generation from noise input.

CNN is widely used in various image recognition tasks [18] for its great capability of achieving significant accuracy. Recent deep learning works in regularization techniques like Batch Normalization [19], Dropout [20], have increased the performance of CNN and other CNN-based architectures significantly. Batch Normalization, Dropout is used in our proposed architecture as a regularization mechanism in both the G and D models to prevent over-fitting and to get a good generalization. Convolutional Neural Network (CNN) based VAE is used in the Generator model (G model). Discriminator model (D model) is a CNN that works as a loss function for the G model.

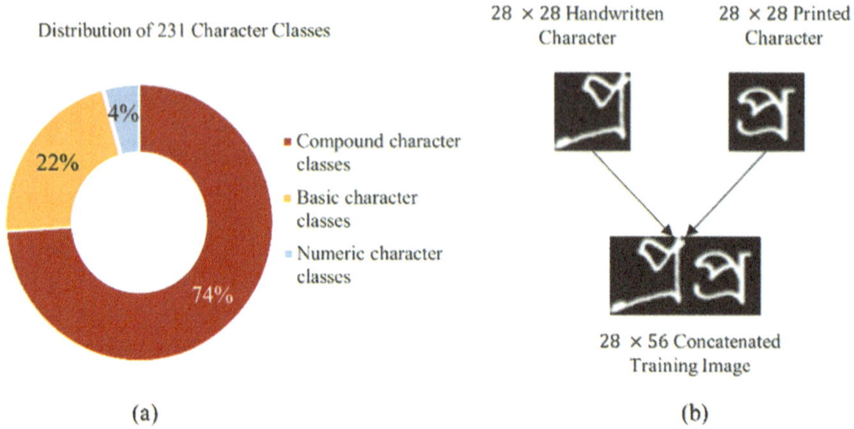

Fig. 1 **a** Distribution of classes. 171 compound character classes, 50 basic character classes, and 10 numeral classes are taken into consideration in this work. **b** Preparation of training dataset. A handwritten character image and a printed character image were concatenated before using in generator model for training purposes

3 Dataset

In this work, we have used CMATERdb [21] database that is publicly available and the biggest dataset of Bangla handwritten characters [22]. It is widely used in many researches [23]. This dataset contains a total of 59,922 handwritten Bangla character images of 231 classes (171 compounds, 50 basic, 10 numeric). We have also created a target printed character dataset that contains printed character images of 231 classes. The images of both of the datasets were converted to 28 × 28 dimensions. Some of the handwritten character images were concatenated with it's corresponding printed character images. This attached image was used for training purposes. Some handwritten character images were kept aside for testing the proposed model. Figure 1a shows the distribution of classes. Figure 1b illustrates the preparation of the training dataset.

4 Proposed Architecture for Bangla Printed Character Generation

Variational Generative Adversarial Network (VAE-GAN) is used to generate Bangla printed character (BPC) from Bangla handwritten character (BHC). The model consists of two Deep Convolutional Networks; the Generator Network and the Discriminator Network. Discriminator Network, as well as Generator Network, used Dropout, which indicates the inter-node connection in Generator as well as in Discriminator can be remained silent for a minibatch with a specific probability.

Dropout helps all nodes to train better. When we use dropouts, all nodes do not remain concurrently active in the network that provides every node with the opportunity to train. This also helps to prevent overfitting.

GAN uses the combination of Deep Learning and game theory collectively. In the GAN Generator model (G model) and the Discriminator model (D model) chase against each other. G model learns to produce some image/video/audio, and D model learns to predict which one is natural image/video/audio or which one is artificially generated by G model. Using the feedback of the D model's output, the G model updates its weight in such a way that it is capable of generating contents of the probability distribution of real data distribution (real printed-character). The generated probability distribution is compared with the probability distribution of natural image/audio/video. If they match, the D model has predicted the generated image as real content (not generated).

$$\min_{G} \max_{D} V(D, G) = \mathbb{E}_{\mathrm{X}\ \mathrm{P_{data}(x)}}\left[\log \mathrm{D(x)}\right] + \mathbb{E}_{\mathrm{Z}\ \mathrm{P_{data}(z)}}\left[\log(1 - \mathrm{D(G(z))})\right] \quad (1)$$

D model $D(x)$ which represents the probability of whether data x is from Bangla handwritten-character (BHC) image or Bangla printed-character (BPC) image. Aim of training is the maximization of the probability of predicting correct class to the training sample as well as the sample generated from the G model. G model, $G(z)$ is a differentiable function. G model is trained concurrently so that it can be minimized to $\log(1 - \mathrm{D}(z))$. In a word, G and D play minimax game with each other so that they can reach equilibrium with value function defined as $V(G, D)$.

In this proposed architecture, VAE-GAN is used as an image translator for generating printed-character from handwritten characters. It has been widely understood that the GAN architecture consists of two distinctly trained neural network naming: the D model (Discriminator) and the G model (Generator). In a conceptual overview of the proposed architecture G model takes input a BHC and generates an output image. Generated output image along corresponding BPC is fed into the D model so that it can train and learn the probability distribution of real BPC and generated images. Feedback from the D model is used by G model so that G model can generate images that match the probability distribution function of real BPC. G model and D model chase each other in a way that G model becomes more efficient at generating realistic images, and D model become more efficient at detecting fake images. In this process, they try to find an equilibrium that is optimal for the Generator. A generalized architecture of GAN is shown in Fig. 2.

The G model is based on VAE, which is developed as U-Net architecture that has an encoder-decoder. The G model takes an input of handwritten character. G model produces synthetic images by down-sampling (encoding) input image to bottleneck. This stage can be considered as a feature extractor, and afterwards, the network performs up-sampling (decoding) from the bottleneck to generate the required output image (in our case, the printed letter). U-Net architecture is demonstrated in Fig. 3 where x is input, and y is output.

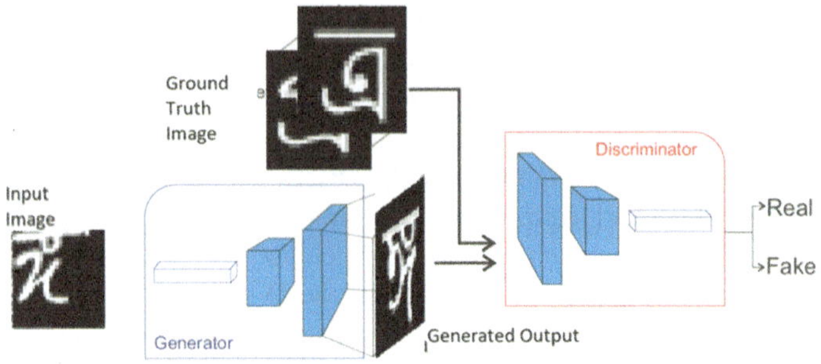

Fig. 2 Generic GAN architecture

Fig. 3 U-Net architecture

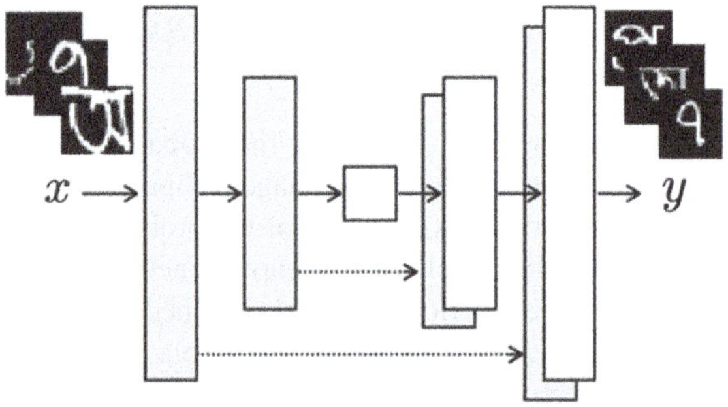

D model: This model has used two input architectures one for the handwritten character and the other for the printed character. Both the handwritten character and the printed character are fed into the model via tensor form $1 \times 28 \times 28 \times 1$. They are concatenated together to produce a tensor of (1, 28, 56, 1).

After that, 3 convolutional units have been used. Each of these convolutional units consists of the same convolution, activation function LeakyReLU [24], Batch Normalization, Dropout, Maxpooling2d. The First convolution unit has two similar convolutional layers, which contain 64 and 128 filters, respectively. Dropout has a probability of 0.5 for each output node along with MaxPooling2D, which causes the image dimension to reduce to half of the dimension of the previous layer. This converts the tensor to (12, 12, 128). The Second convolution unit has only one convolutional layer that contains 256 filters, Batch Normalization, and activation function is LeakyReLU alongside Dropout has 50% node connection off probability along with MaxPooling2D which convert the tensor to 5, 5, 256. The third convolution unit also has only one convolutional layer with 512 filters, Batch Normalization, LeakyReLU activation function, and 50% Dropout along with MaxPooling2D, which converts the image matrix to the half size of the previous.

Flatten is used for vectorization, which has provided a 512-vector size. Then it is fed to a set of fully connected layers. After each fully connected layer, Batch

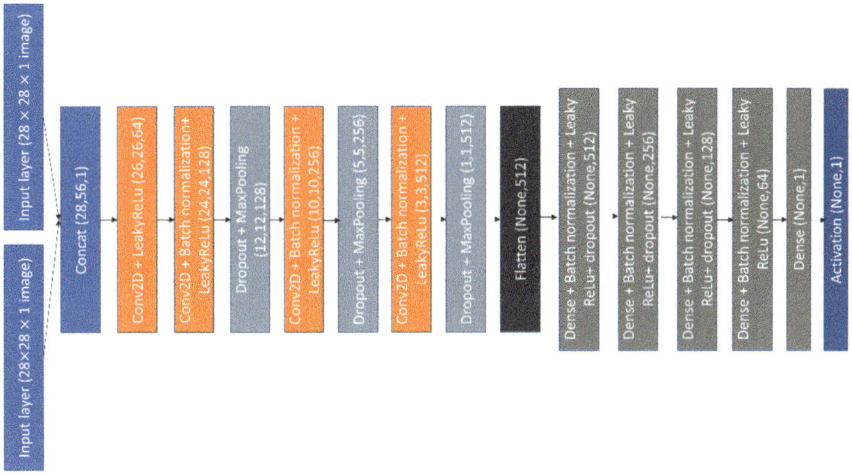

Fig. 4 Overview of the discriminator model

Normalization and Dropout have been used. There are 4 fully connected layers, and each having 256, 128, 64 filters, then the final layer is 1 neuron, and final layer activation function is sigmoid because D model is binary output NN which distinguishes an image whether handwritten or printed character. Figure 4 illustrates the discriminator model.

G Model: In the G model, VAE has been used as a Generator that has encoder and decoder. The encoder part is used for down-sampling and bottleneck, whereas the decoder part is used for up-sampling from the bottleneck, in other words, image generation.

There are five convolutional units, each having convolution layer, batch-normalization, Dropout of 25% probability, and LeakyReLU as activation function is used in down-sampling (Encoding). 128, 256, 512, 1024, and 2048 convolutional filters have been used respectively, starting from the first to the fifth convolutional unit. The fifth layer contains 2048 convolutional filters, and this layer works as a bottleneck for the G model (Generator).

In up-sampling (Decoding) there are four convolutional units, each having ConvTranspose2D layer, Batch Normalization, Dropout with a probability of 25% and LeakyReLU is used as an activation function. 2048, 1024, 512 and 1 is used as filter respectively in our up-sampling starting from the first to the fourth up-sampling unit. The first up-sampling has 2048 filters since there are similar numbers of filters in the bottleneck. The last or final unit contains just one filter that is for generating grayscale images as our output. Up-sampling is done in such a way that image can match the 28×28 matrix size with the final layer. The activation function for the final layer is tanh, and this function produces an output within a range of -1 to 1. This activation function helps in image generation. Figure 5 provides an overview of the generator model.

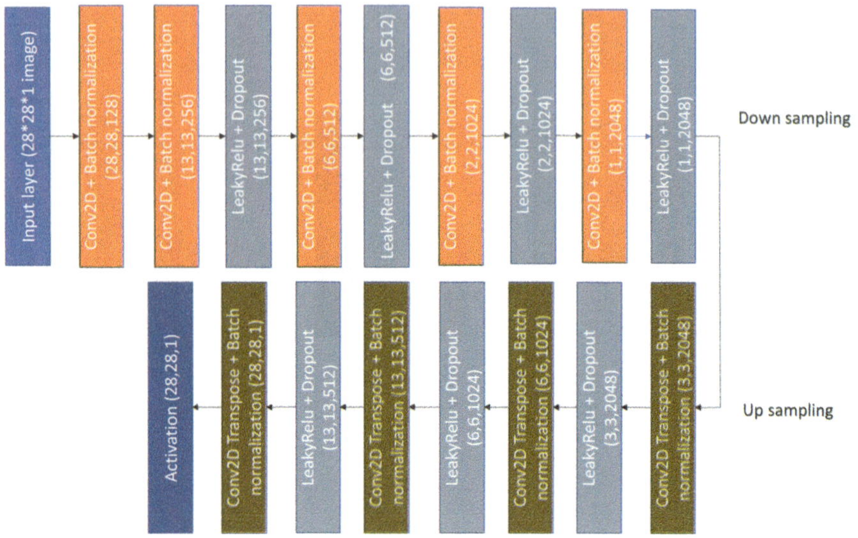

Fig. 5 Overview of generator model

Training: In the training process, the Discriminator Neural Network is trained on a single batch of ground truth image subsequently, the Generator generates a single batch of the printed image, and finally the GAN is updated by the Discriminator feedback.

Figure 6 demonstrates the process of training of the GAN model. During the training period, the Discriminator model tries to beat the Generator model and vice-versa.

The Generator continuously tries to upgrade the probability distribution by feedback form D model so that generated image can match the probability density of the real target image. On the other hand, the D model learns to distinguish between realistic image and generated images. D model works as a loss function for the G model. The training goal is to find equilibrium so that loss of the G model can be minimized, and loss of D model can be maximized.

In initial epochs, the GAN loss remains quite high. As the epochs advance, GAN loss decreases, and the generated image starts to eventually develop as similar to that of the Ground Truth images. At Epoch 1, almost nothing has been generated. After 40 epochs, it starts to bear similarities, and after 70 epochs, it becomes clearer and more similar to that of ground truth image. After 70 epochs, GAN loss is significantly low compared to the first epoch, and it will decline slowly with more epochs. But after specific epochs, the rate of decreasing the loss will be very low.

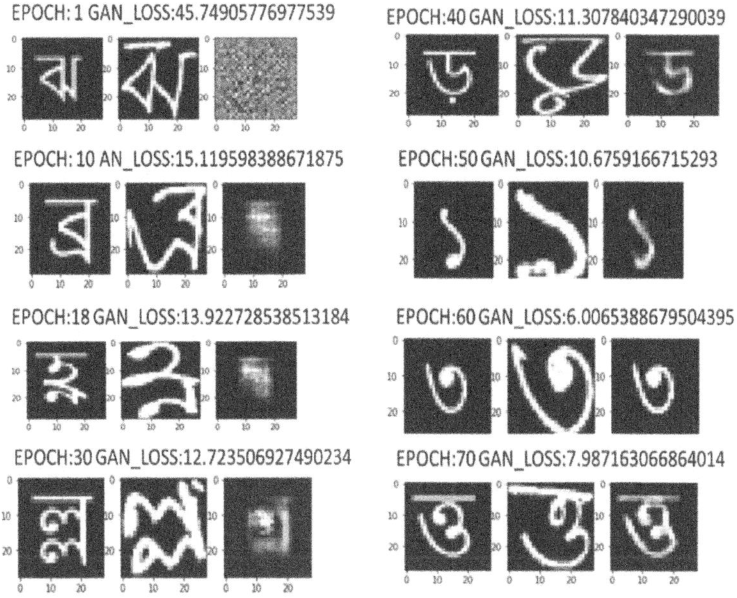

Fig. 6 GAN training

5 Experimental Result

In the training process, there is three loss functions Discriminator loss for the real image, Discriminator loss for the fake image, GAN loss. Discriminator loss for a real image means loss when the real image is into D model. Discriminator loss for a fake image means loss when the fake (generated) image is fed into D mode. GAN loss means the difference between data distribution of the generated image and the target image.

GAN has been trained for 100 epochs. Figure 7 shows the loss value of the G model and D model with respect to epochs. The discriminator loss for real image (handwritten) is high in the first iteration. The loss gradually reduces as the training starts to feed handwritten characters, and the Neural Network consistently updates its weight from the first discriminator loss. This loss consistently reduces over time.

Discriminator loss for the fake image is initially low as the data distribution does not match with the D model distribution. When Generator learns the data distribution from Discriminator's real loss, the fake loss increases. Initially, the GAN loss is quite high and reduces over time. Training will lead to the point that the rate of decreasing GAN loss will be very low. That point is a training goal where the D model and G model will be in equilibrium. GAN loss will never become zero, which is expected, but it is not possible. It shows the diversity of data distribution of output image as the output will be contained some noise and never be exactly like the target image.

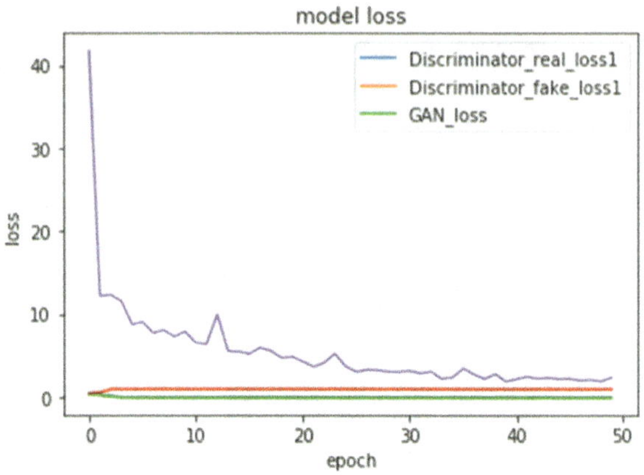

Fig. 7 Loss graph for discriminator and generator

This diversity leads to data augmentation, and because termination of GAN training can be chosen so that preferred data augmentation for OCR training can be controlled. On the other hand, if GAN loss can be made low enough by train, then generated character can be used as a printed-character. However, there will be some noise in the generated image. The human eye will be capable enough to detect the difference between generated and original images.

After completing 100 epochs, the generated images are very likely to the ground truth images. Figure 8 shows a few examples of generated printed letters from their respective handwritten character. Input data are not used in the training process they are from different data source but preprocessed to 28 × 28 × 1 pixel size so that images can be fed into Generator. In Fig. 8 it is quite clear that all the input image matches the output image maybe they are distorted and have some noise but the output image and input image belong to the same class. This defines that out GAN model can correctly classify class labels.

Some generated synthetic images are presented in Fig. 9. Figure 9 shows that some images are distorted because of noise, which was introduced due to GAN loss. GAN loss can be control by choosing termination of GAN training. The time duration of GAN training will depend on the use of the Generator. If we want to use Generator as printed character generator, then training time should be long so that GAN loss can minimize enough. On the other hand, if we want to use the Generator as a data generator for the OCR training, GAN should be trained in a controlled manner so that there remain GAN loss. If the GAN loss is high, generated data will have more noise. If the GAN loss is low, generated data will have low noise.

It can be noticed that the above example displays some of the synthetic images, which are distorted, and some are not. Distortion typically takes place due to the curvature present in Bangla handwritten characters. In Bangla script, there are many characters which appear to be very similar to each other. For instance, "ত" and "ভ", "ও" and "ও"and many more look similar. Besides, the same character may also possess different forms. Curvature in handwritten characters may vary from person to person, and this affects the generalized form of curvature for

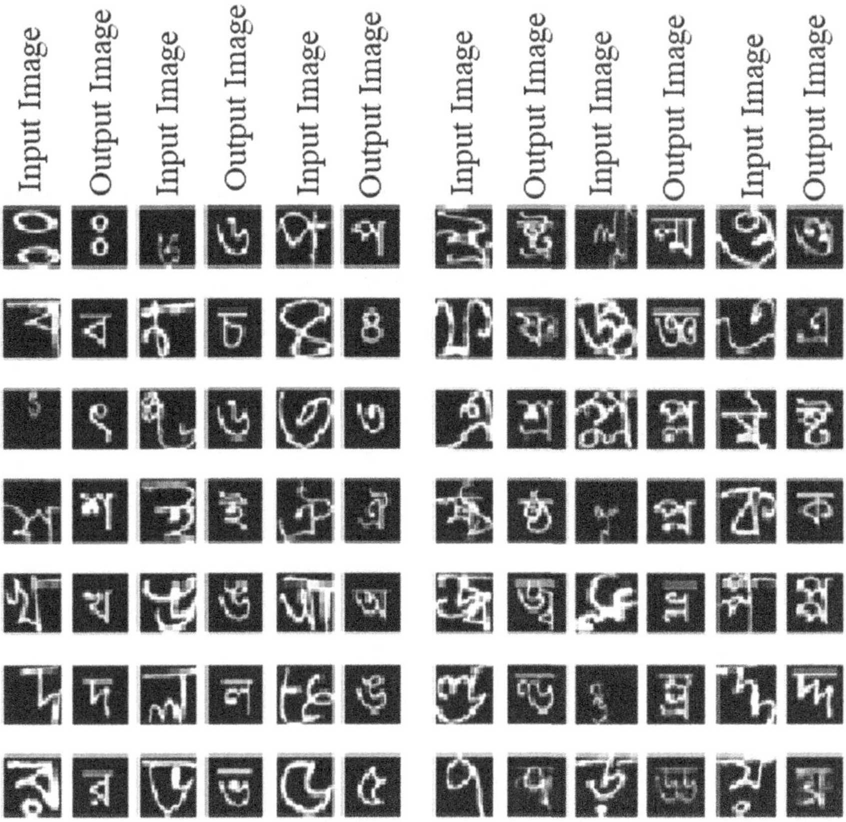

Fig. 8 Example of generated printed letter for respective handwritten character

a specific script. As Bangla script looks very similar in printed form, the curvature variations in handwritten character have overlapped with different scripts in many cases. Hence, the generated images are sometimes distorted or overlapped.

6 Conclusion

Generative adversarial networks offer to learn deep representations devoid of widely annotated training of data. This is achieved by a pair of Neural Networks that involve a competitive process of deriving feedback. The Generator NN tries to outperform the Discriminator NN by generating images close to the ground truth images and efforts at minimizing its loss. Nevertheless, only by increasing the epochs will possibly not guarantee the generation of better images. If the rate of learning is too low, the Discriminator starts to win and the quality of the images keeps on distorting. Adding more layers to the Generator can be an approach to improve the model.

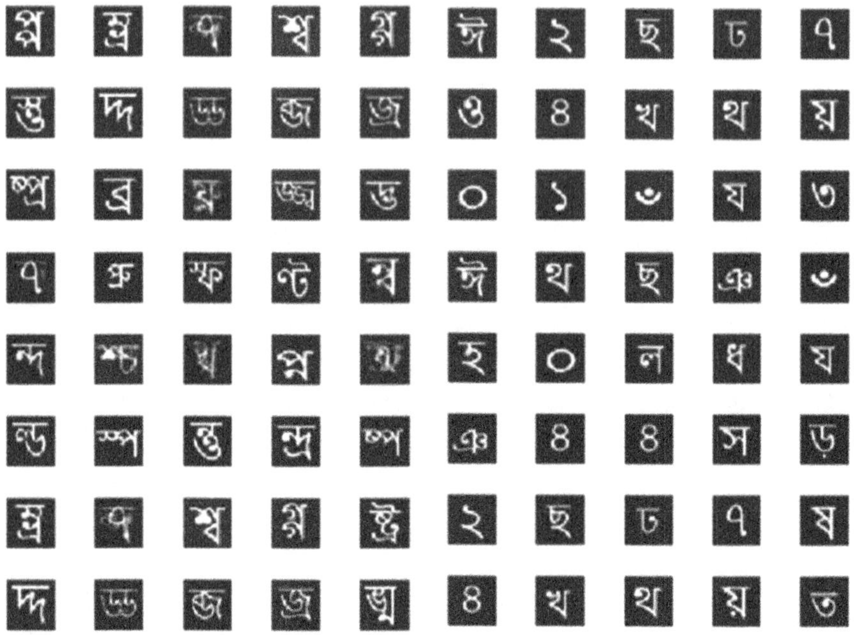

Fig. 9 Example of some generated synthetic images

GAN is commonly applied to generate images of Google maps or cartoon generation, image to image translation, etc. This work is based on Generative Adversarial Network to reconstruct Bangla characters. Bangla handwritten character recognition and conversion of handwritten character to typed character format is the first step towards establishing an automated handwritten document to typed document conversion system. Digitization of handwritten documents in a typed format from a handwritten document is much necessary to store essential documents. In addition to this, it can make things easier for elderly people who are not acquainted with the computer. The proposed model can generate around 231 handwritten characters in the Sutoni font. This can be used further for formatted document generation.

References

1. Goodfellow, I., Pouget-Abadie, J., Mirza, M., et al.: Generative adversarial nets. In: Advances in Neural Information Processing Systems, pp. 2672–2680 (2014)
2. Kingma, D.P., Welling, M.: Auto-encoding variational bayes. arXiv Prepr arXiv13126114 (2013)
3. Radford, A., Metz, L., Chintala, S.: Unsupervised representation learning with deep convolutional generative adversarial networks. arXiv Prepr arXiv151106434 (2015)
4. Karras, T., Aila, T., Laine, S., Lehtinen, J.: Progressive growing of gans for improved quality, stability, and variation. arXiv Prepr arXiv171010196 (2017)

5. Reed, S., Akata, Z., Yan, X., et al.: Generative adversarial text to image synthesis. arXiv Prepr arXiv160505396 (2016)
6. Xu, T., Zhang, P., Huang, Q., et al.: Attngan: fine-grained text to image generation with attentional generative adversarial networks. In: Proceedings of the IEEE Conference on Computer Vision and Pattern Recognition, pp. 1316–1324 (2018)
7. Lu, Y., Wu, S., Tai, Y.-W., Tang, C.-K.: Image generation from sketch constraint using contextual gan. In: Proceedings of the European Conference on Computer Vision (ECCV), pp. 205–220 (2018)
8. Isola, P., Zhu, J.-Y., Zhou, T., Efros, A.A.: Image-to-image translation with conditional adversarial networks. In: Proceedings of the IEEE Conference on Computer Vision and Pattern Recognition, pp. 1125–1134 (2017)
9. Wang, T.-C., Liu, M.-Y., Zhu, J.-Y., et al.: High-resolution image synthesis and semantic manipulation with conditional gans. In: Proceedings of the IEEE Conference on Computer Vision and Pattern Recognition, pp. 8798–8807 (2018)
10. Zhu, J.-Y., Park, T., Isola, P., Efros, A.A.: Unpaired image-to-image translation using cycle-consistent adversarial networks. In: Proceedings of the IEEE International Conference on Computer Vision, pp. 2223–2232 (2017)
11. Sangkloy, P., Lu, J., Fang, C., et al.: Scribbler: Controlling deep image synthesis with sketch and color. In: Proceedings of the IEEE Conference on Computer Vision and Pattern Recognition, pp. 5400–5409 (2017)
12. Liu, Y., Qin, Z., Luo, Z., Wang, H.: Auto-painter: Cartoon image generation from sketch by using conditional generative adversarial networks. arXiv Prepr arXiv170501908 (2017)
13. Nazeri, K., Ng, E., Ebrahimi, M.: Image colorization using generative adversarial networks. In: International Conference on Articulated Motion and Deformable Objects, pp. 85–94 (2018)
14. Han, C., Hayashi, H., Rundo, L., et al.: GAN-based synthetic brain MR image generation. In: 2018 IEEE 15th International Symposium on Biomedical Imaging (ISBI 2018), pp. 734–738 (2018)
15. Jeng, B.-S., Chang, K.-Y., Liu, T.-S., et al.: An automatic Chinese character maltifont generating system. In: Applications of Digital Image Processing IX, pp. 181–186 (1986)
16. Li, M., Wang, J., Yang, Y., et al.: Improving GAN-based calligraphy character generation using graph matching. In: 2019 IEEE 19th International Conference on Software Quality, Reliability and Security Companion (QRS-C), pp. 291–295 (2019)
17. Nishat, Z.K., Shopon, M.: Synthetic class specific Bangla handwritten character generation using conditional generative adversarial networks. In: 2019 International Conference on Bangla Speech and Language Processing (ICBSLP), pp. 1–5 (2019)
18. Al Banna, M.H., Haider, M.A., Al Nahian, M.J., et al.: Camera model identification using deep CNN and transfer learning approach. In: 2019 International Conference on Robotics, Electrical and Signal Processing Techniques (ICREST), pp. 626–630 (2019)
19. Ioffe, S., Szegedy, C.: Batch normalization: accelerating deep network training by reducing internal covariate shift. arXiv Prepr arXiv150203167 (2015)
20. Srivastava, N., Hinton, G., Krizhevsky, A., et al.: Dropout: a simple way to prevent neural networks from overfitting. J. Mach. Learn. Res. **15**, 1929–1958 (2014)
21. Das, N., Acharya, K., Sarkar, R., et al.: A benchmark image database of isolated Bangla handwritten compound characters. Int. J. Doc. Anal. Recogn. **17**, 413–431 (2014)
22. Ghosh, T., Chowdhury, S.M., Yousuf, M.A., et al.: A comprehensive review on recognition techniques for Bangla handwritten characters. In: 2019 International Conference on Bangla Speech and Language Processing (ICBSLP), pp. 1–6 (2019)
23. Ghosh, T., Abedin, M.M.-H.-Z., Chowdhury, S.M., et al.: Bangla handwritten character recognition using MobileNet V1 architecture. Bull. Electr. Eng. Informatics **9**, 2547–2554 (2020)
24. Xu, B., Wang, N., Chen, T., Li, M.: Empirical evaluation of rectified activations in convolutional network. arXiv Prepr arXiv150500853 (2015)

The manufacturer's authorised representative in the EU is Springer Nature Customer Service Centre GmbH, Europaplatz 3, 69115 Heidelberg, Germany. If you have any concerns regarding our products, please contact ProductSafety@springernature.com

Printed and bound by CPI Group (UK) Ltd, Croydon, CR0 4YY
30/09/2026
02225710-0009